CREATION COLLECTION

# HUMAN ORIGINS

Jeffrey P. Tomkins
with Frank Sherwin, Brian Thomas,
and Timothy Clarey

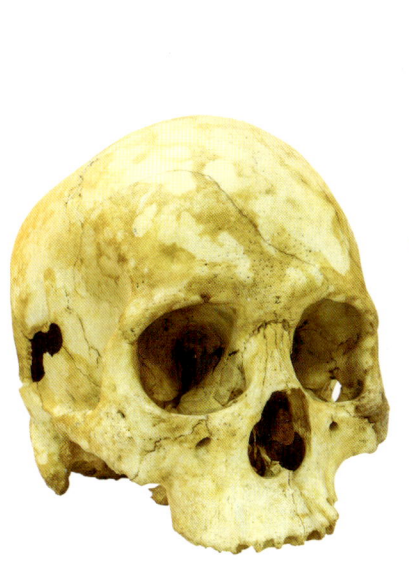
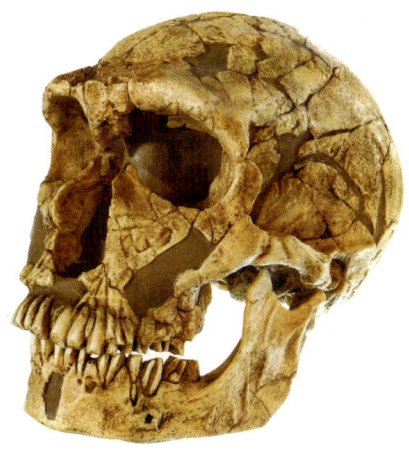

CREATION COLLECTION

# HUMAN ORIGINS

Jeffrey P. Tomkins
with Frank Sherwin, Brian Thomas,
and Timothy Clarey

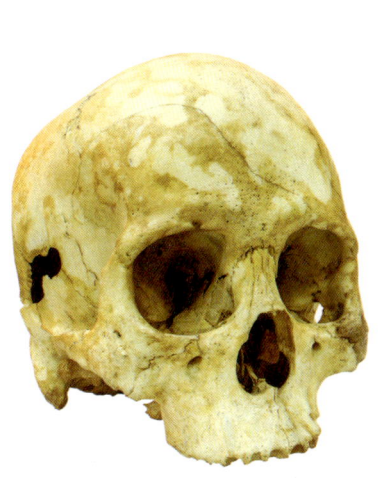 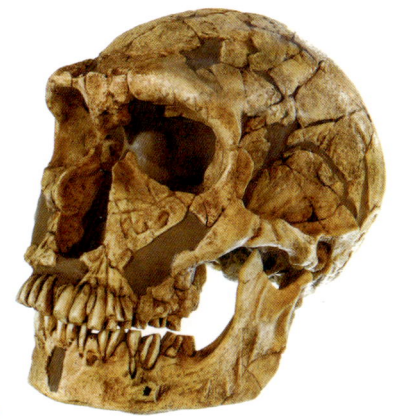

Dallas, Texas
ICR.org

# Human Origins

by Jeffrey P. Tomkins, Ph.D., with Frank Sherwin, D.Sc. (Hon.), Brian Thomas, Ph.D., and Timothy Clarey, Ph.D.

First Printing: December 2022

Copyright © 2022 by the Institute for Creation Research. All rights reserved. No portion of this book may be used in any form without written permission of the publisher, with the exception of brief excerpts in articles and reviews. For more information, write to Institute for Creation Research, P. O. Box 59029, Dallas, TX 75229.

All Scripture quotations are from the New King James Version.

ISBN: 978-1-957850-11-5
Library of Congress Control Number: 2022948622

Please visit our website for other books and resources: ICR.org

Printed in the United States of America.

# TABLE OF CONTENTS

Introduction ..................................................................... 7
1. A Literal Adam Is a Gospel Issue .............................. 9
2. Separate Studies Converge on Human-Chimp DNA Dissimilarity ...................................................... 13
3. How Do Hominids Fit with the Bible? ................... 17
4. 95% of the Human Genome Can't Evolve ............. 23
5. Scientists Discover Another Function of "Junk DNA" ........................................................... 27
6. Ape and Human Spit Are Radically Different ...... 31
7. *Australopithecus* Ate Like an Ape ........................... 35
8. Codons Are Functional After All ............................ 39
9. Denisovan Epigenetics Reveals Human Anatomy .... 43
10. Denisovan DNA Shown to Be Human…Again ..... 47
11. DNA Science Disproves Human Evolution .......... 51
12. Does Recent Research Support Human Evolution? ................................................................. 59
13. Fossil Ape Skull Is a Game Ender .......................... 65
14. Recent Humans with Archaic Features Upend Evolution ..................................................................... 69
15. *Homo erectus*: The Ape Man That Wasn't ............ 73
16. *Homo naledi*: Another Failed Evolutionary Ape Man ..................................................................... 83
17. Human Chromosome 2 Fusion Never Happened ... 97
18. Human Genome 20th Anniversary—Junk DNA Hits the Trash ........................................................... 109
19. Human High-Altitude Habitation Reveals Adaptive Design ...................................................... 113
20. Humans and Neanderthals Are More Similar than Polar and Brown Bears .................................. 117
21. Lucy Languishes as a Human-Ape Link ............. 121
22. Man: Smart from the Start ..................................... 129

## INTRODUCTION

The special creation of humanity plays an essential role in the gospel. Evolution claims we are the result of eons of death and evolutionary change. But Romans 5:12 states that death only entered the world when the first humans sinned. And Genesis 1 makes it clear that God made the first man and woman in His image on the sixth day of creation. Jesus confirms this truth in Matthew 19:4: "Have you not read that He who made them *at the beginning* made them male and female?"

In this book, ICR geneticist Dr. Jeffrey P. Tomkins shares his extensive research on the human genome and shows that we didn't evolve from an ape-like ancestor. Our genome is not full of "junk DNA" but is rather fully functional. As Dr. Tomkins says in the first chapter:

> The idea that humans somehow emerged from apes—after billions of years of primeval Earth history, followed by millions of years of evolution—is unbiblical and unsupported by sound science. There is no need for Christians to compromise on this important issue.

Other ICR scientists join Dr. Tomkins to demonstrate how science confirms the biblical message and the divine engineering of our Creator Jesus Christ. We can trust the Bible about our origins.

# 1

## A LITERAL ADAM IS A GOSPEL ISSUE

Jeffrey P. Tomkins, Ph.D.

- Many Christians who accept evolution think a literal Adam is irrelevant because it isn't a gospel issue.
- This is not only unbiblical, it's unnecessary, since the science doesn't support human evolution.
- Sin and death entered the world through Adam, which is why Jesus came to save us.
- The Bible affirms the historicity and necessity of a literal Adam and Eve.

Did Adam really exist? Does it matter? Many academics and even Christians claim that humans descended from apes through an evolutionary process over millions of years. This claim contradicts the biblical account of mankind's unique creation in God's image about 6,000 years ago. A key problem with the evolutionary position is that the fossil record contains no evidence of a transition from apes to humans.

*Australopithecus* are ape-like fossils thought to represent the first precursor to the genus *Homo*, or human. However, nothing has been found to bridge the gap between the two groups. In a 2016 Royal Society paper titled "From *Australopithecus* to *Homo*: the transition that wasn't," two conventional paleontologists state:

> Although the transition from *Australopithecus* to *Homo* is usually thought of as a momentous transformation, the fossil record bearing on the origin and earliest evolution of *Homo* is virtually undocumented.[1]

Even the field of human-ape DNA similarity research has come up empty in this regard. Both creationists and evolutionists documented that the human and chimp genomes are no more than 85% similar.[2] Humans and chimps must be 98-99% similar to have evolved from a common ancestor over three to six million years. The scientific data from both paleontology and genetics demonstrate a chasm of discontinuity between humans and apes. This situation is clearly on the side of the Bible's account of human history.

*Australopithecus africanus,* Taung child

Many Christians think they should not be overly concerned about a literal Adam because it's not directly related to the gospel of Jesus Christ. But this

10

is false. Not only are humans created uniquely in the image of God, but the story of a historical Adam is foundational to the gospel.

Sin entered the picture through a literal Adam and Eve, along with death, misery, and corruption. This curse accounts for the central problem of evil in the world. Romans 5:12 says, "Therefore, just as through one man sin entered the world, and death through sin, and thus death spread to all men, because all sinned." This foundational gospel truth is repeated in 1 Corinthians 15:22: "For as in Adam all die, even so in Christ all shall be made alive." And the pervasive and disastrous effect of mankind's sin on the whole creation is stated in Romans 8:21-22: "Because the creation itself also will be delivered from the bondage of corruption....For we know that the whole creation groans and labors with birth pangs together until now."

Jesus Christ affirmed the historicity of a literal human couple during His earthly ministry. In Matthew 19:4, He tells us, "Have you not read that He who made them at the beginning made them male and female." The Lord confirmed the Genesis account of humanity's creation and affirmed that this occurred at the very beginning of Earth's time frame. And we know from detailed genealogies throughout the Bible combined with scriptural data on times of birth and death[3] that Earth is approximately 6,000 years old.

The idea that humans somehow emerged from apes—after billions of years of primeval Earth history, followed by millions of years of evolution—is unbiblical and unsupported by sound science. There is no need for Christians to compromise on this important issue. Adam was a real person.

**References**

1. Kimbel, W. H. and B. Villmoare. 2016. From *Australopithecus* to *Homo*: the transition that wasn't. *Philosophical Transactions of the Royal Society B*. 371 (1698): 20150248.
2. Tomkins, J. P. 2018. Separate Studies Converge on Human-Chimp DNA Dissimilarity. *Acts & Facts*. 47 (11): 9.
3. Johnson, J. J. S. 2008. How Young Is the Earth? Applying Simple Math to Data in Genesis. *Acts & Facts*. 37 (10): 4.

# 2

# SEPARATE STUDIES CONVERGE ON HUMAN-CHIMP DNA DISSIMILARITY

Jeffrey P. Tomkins, Ph.D.

- Advances in genetic research continue to bolster the genetic case against evolution.
- Both secular and creation genetics research confirm the same percentage of human-chimp DNA dissimilarity.
- No more than about a 1% DNA difference between humans and chimps is required to make human evolution plausible, so 15% is too great a discrepancy to ignore.

The improvement of DNA sequencing technology, along with scientific advances in the field of genomics, is proving to be a profound enemy of evolution. Two discoveries that challenge the human

evolution paradigm were reported nearly simultaneously—one by a secular scientist and the other by myself. Remarkably, the corroborating data produced in both reports are in perfect agreement.

A key element of the evolutionary paradigm is the idea that human DNA and chimpanzee DNA are 98.5% identical. This level of DNA similarity is needed to undergird the hypothesis that humans and chimps shared a common ancestor three to six million years ago. Based on known mutation rates in both humans and chimps, anything significantly less than a 98.5% DNA similarity would destroy the foundation of the entire theory.

When I began studying the scientific literature on the subject, I realized there were serious problems with the evolutionary idea of nearly identical human and chimp DNA. In every publication I evaluated, it became clear researchers had cherry-picked highly similar DNA sequences that supported evolution and discarded the data that were dissimilar.[1] I recalculated DNA similarities from these studies by factoring back in omitted data and obtained drastically lower levels of human-chimp DNA similarity—between 66% and 86%.

Another major issue I uncovered is that the chimpanzee genome was literally put together to resemble the human genome.[2] This little-known fact was accomplished by taking the small snippets of DNA produced after sequencing and lining them up on the human genome. The human genome guided the researchers throughout the chimp genome assemblage process.

Despite these issues, brought about because of scientists' evolutionary bias, continuing improvement in DNA sequencing technology is slowly bringing the truth to light. A more recent version of the chimpanzee genome was completed, and the results not only validate my past research but also spectacularly confirm research I published.[3]

The research paper for the new chimp genome completely sidesteps the issue of DNA similarity with humans.[4] Nevertheless, University of London evolutionist Richard Buggs analyzed the results of a comprehensive comparison of the new chimp genome with the human one and posted his shocking antievolutionary findings. He stated, "The percentage of nucleotides in the human genome that had one-to-one exact matches in the chimpanzee genome was 84.38%."[5]

What makes Dr. Buggs' analysis more amazing is the fact that my own published research using a different algorithm gave the same results. In my study, I aligned 18,000 random pieces of high-quality chimp DNA about 31,000 DNA letters long (on average) onto human and several different versions of the chimp genome. Not only did my data show that the older version of the chimp genome (PanTro4) that had been used to support evolution was deeply flawed and humanized, but they also showed the aligned segments of chimp DNA were on average only 84.4% identical to human—the same level reported by Buggs.

These results by both myself and Buggs also closely match a 2016 study I published that indicated the overall human-chimp DNA similarity was likely no more than 85%.[6] Based on the more recent research, the difference between the human and chimp genomes is estimated to be about 15%.

A 15% DNA difference between humans and chimps is a discrepancy that can't be ignored when no more than about a 1% difference is required to make human evolution seem at all plausible. Once again, the scientific accuracy of the Bible is vindicated regarding the uniqueness of humans as stated in Genesis 1:27: "So God created man in His own image; in the image of God He created him; male and female He created them."

*References*

1. Tomkins, J. and J. Bergman. 2012. Genomic monkey business—estimates of nearly identical human-chimp DNA similarity re-evaluated using omitted data. *Journal of Creation*. 26 (1): 94-100.
2. Tomkins, J. 2011. How Genomes Are Sequenced and Why It Matters: Implications for Studies in Comparative Genomics of Humans and Chimpanzees. *Answers Research Journal*. 4: 81-88.
3. Tomkins, J. 2018. Comparison of 18,000 De Novo Assembled Chimpanzee Contigs to the Human Genome Yields Average BLASTN Alignment Identities of 84%. *Answers Research Journal*. 11: 215-219.
4. Kronenberg, Z. N. et al. 2018. High-resolution comparative analysis of great ape genomes. *Science*. 360 (6393): eaar6343.
5. Buggs, R. How similar are human and chimpanzee genomes? Posted on richardbuggs.com July 14, 2018, accessed August 9, 2018.
6. Tomkins, J. 2016. Analysis of 101 Chimpanzee Trace Read Data Sets: Assessment of Their Overall Similarity to Human and Possible Contamination With Human DNA. *Answers Research Journal*. 9: 294-298.

# 3
# HOW DO HOMINIDS FIT WITH THE BIBLE?
Brian Thomas, Ph.D.

- Most conventional scientists believe humans evolved from ape-like creatures that lived millions of years ago.
- If this is true, then the Bible is wrong when it describes the creation of humanity only thousands of years ago.
- All fossils that conventional scientists claim are human ancestors fall into three categories: ape, imaginary, or human.
- These ape and human fossils fit the Bible's narrative.

A survey showed that the most persuasive argument for evolution comes from the iconic drawing of the apes-to-man parade.[1] This popular picture illustrates ape-like animals evolving into a human. If this image reflects actual history, then the account in Genesis is wrong. If we came from apes, then we didn't come from Adam and Eve. That also casts doubt on the other Scriptures—and their human authors—that refer to Adam as our real ancestor.[2] Do certain fossils demand we take scissors to our Bibles?

Scientific literature and popular media portray

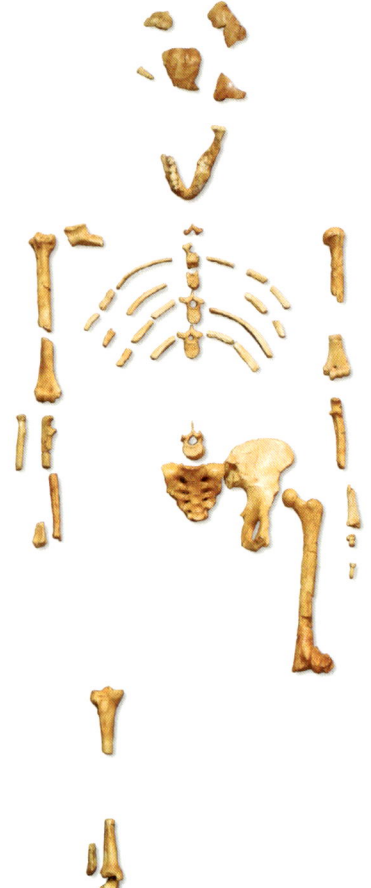

Lucy

hominids as human-like apes or ape-like humans on their way to becoming real humans. Over the last 50 years of looking into human origins, ICR scientists found that hominid fossils fit three creation-friendly groups that leave the biblical Adam intact.

The first group is ape. The fossil nicknamed Lucy is probably the most famous so-called "hominid" in this category. Lucy had locking wrists and ape-like fingers, arms, and ribs, with ape jaws and teeth. Why not call her kind apes? Donald Johanson, who discovered Lucy, claims that Lucy's kind was becoming human. But others, like the late Baron Solly Zuckerman, saw her kind as mere apes with no human ancestry.[3] These fossils rightly bear the ape name *Australopithecus*.

Conventional scientists often present these apes as appearing very human-like. In the 1970s, evolutionists assigned fossilized *human* tracks at Laetoli, Tanzania, to Lucy's kind. That was easier when scientists didn't have enough australopith foot fragments to figure out what its feet looked like. Since then, scientists have found australopiths with feet, and these new fossils confirm its ape grouping by showing it had hands for feet just like chimps do.[4] Does this extinct ape pose a threat to Genesis history? No. Evolution needs natural processes to create new creature kinds, not kill off old ones.

Another candidate named *Australopithecus sediba* fits the second group: imaginary. When the fossil was first described in 2011, paleoanthropologist Darryl DeRuiter said, "This is what evolutionary theory would predict, this mixture of Australopithecene and *Homo*….It's strong confirmation of evolutionary

*Australopithecus sediba*

theory."[5] But other scientists took a closer look and found the real reason for "this mixture." Sediba did not combine different *features* but different *species*. Like the famous Piltdown forgery,[6] Sediba belongs to an imaginary group.[7]

The third group is human. Which fossils belong here? Sometimes it's difficult to tell. Healthy human heads can take many different shapes and sizes.

*Homo floresiensis* (nicknamed "Hobbit") presented a challenge. The skull and other fragments from a remote island in Indonesia came from a small person with a tiny head. Initial reports declared it a possible ancestor, but later work showed an excellent match between Hobbit and people today who have a disease called microcephaly.[8,9] Diseased humans don't show evolution. They show sin's curse on creation.

So far, no fossil fits human evolution. Whether ape, imaginary, or human, fossils confirm created kinds and Adam in our not-so-distant past.

**References**

1. Biddle, D. A. and J. Bergman. 2017. Strategically dismantling the evolutionary idea strongholds. *Journal of Creation*. 31 (1): 116-119.

2. The complete list includes Genesis, Deuteronomy, Joshua, 1 Chronicles, Job, Ezekiel, Luke, Romans, 1 Corinthians, 1 Timothy, and Jude.

3. Roger Lewin wrote about Zuckerman: "'They are just bloody apes,' he is reputed to have observed on examining the australopithecine remains in South Africa." Lewin, R. 1987. *Bones of Contention*. New York: Touchstone, 165.

4. DeSilva, J. M. et al. 2018. A nearly complete foot from Dikika, Ethiopia and its implications for the ontogeny and function of *Australopithecus afarensis*. *Science Advances*. 4 (7): eaar7723.

5. Potter, N. Evolutionary 'Game Changer': Fossil May Be Human Ancestor. *ABC News*. Posted on abcnews.go.com September 8, 2011, accessed February 11, 2020.

6. To manufacture Piltdown man as an evolutionary ancestor, someone filed down and stained an ape jaw to make it fit a human cranium.
7. Ann Gibbons wrote, "'The best candidate' for the immediate ancestor of our genus *Homo* may just be a pretender." Gibbons, A. A famous 'ancestor' may be ousted from the human family. *Science*. Posted on sciencemag.org April 23, 2017, accessed February 11, 2020.
8. Hershkovitz, I., L. Kornreich, and Z. Laron. 2007. Comparative Skeletal Features Between *Homo floresiensis* and Patients With Primary Growth Hormone Insensitivity (Laron Syndrome). *American Journal of Physical Anthropology*. 134 (2): 198-208.
9. Martin, R. D. et al. 2006. Flores Hominid: New Species or Microcephalic Dwarf? *The Anatomical Record Part A*. 288A (11): 1123-1145.

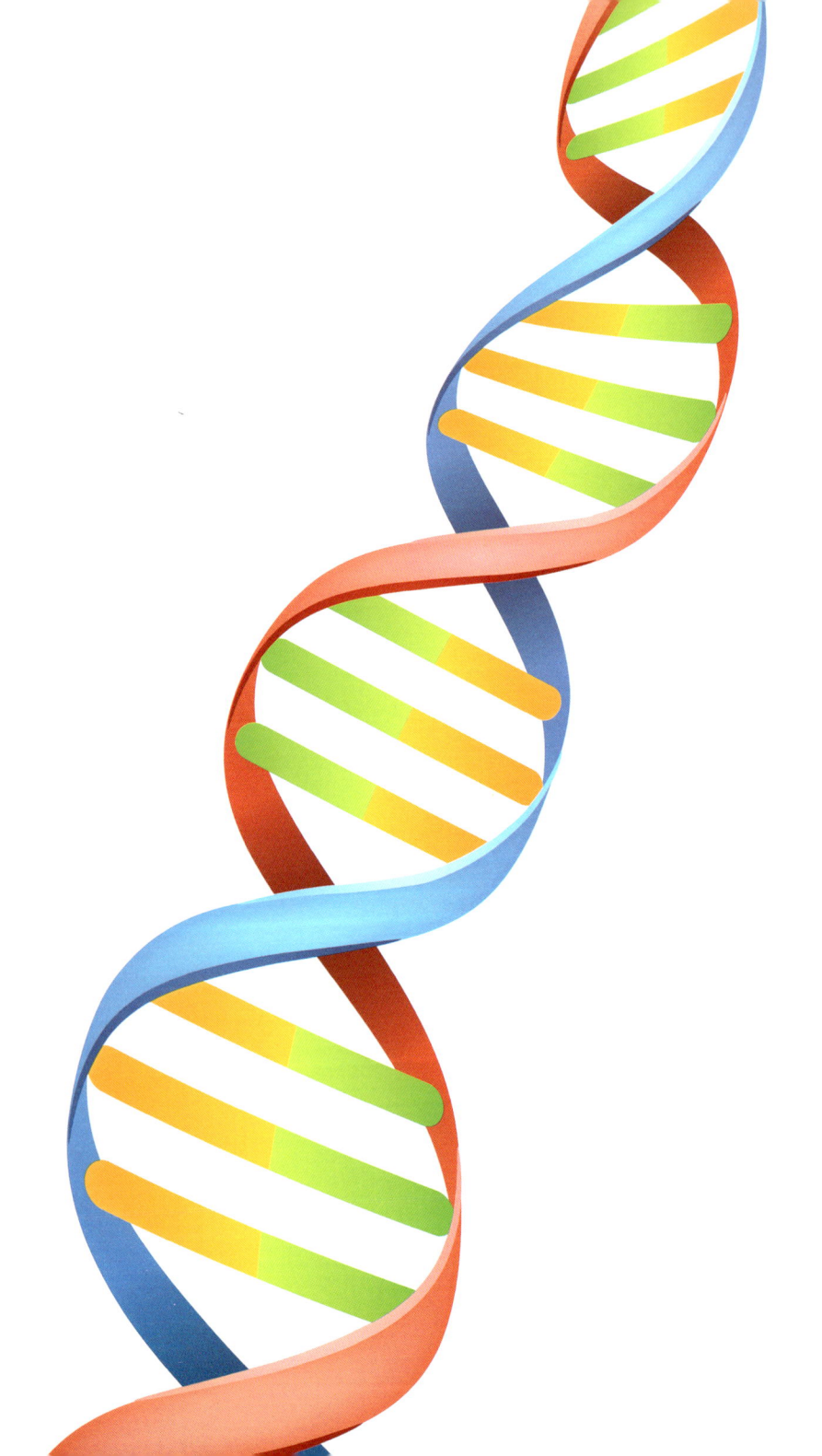

# 4

# 95% OF THE HUMAN GENOME CAN'T EVOLVE

Jeffrey P. Tomkins, Ph.D.

In 2018, a study was released that analyzed vast amounts of data from human genome samples worldwide.[1] Based on the evolutionists' own theoretical model of evolution, 95% of the human genome is "restrained"—it can't evolve.

According to the popular neutral model of evolutionary theory, much of the human genome is nothing but randomly evolving junk. This so-called "neutral DNA" is allegedly not under any "selective restraint" and only serves as material for functional new genes and traits to evolve.

However, in 2012, a vast global consortium of biomedical geneticists working on the ENCODE project released their results. These scientists are more interested in curing human disease than speculative and unproductive research about evolution, and they reported that at least 80% of the human genome demonstrates biochemical function.[2] This is far more function than evolutionary models predicted.

Nevertheless, vocal theoretical evolutionists pushed back and published a variety of papers. They essentially used evolution to prove evolution. As a result of their theoretical calculations based on the premise of evolution, they claimed that the human genome could be no more than 8.2% functional—despite the avalanche of empirical evidence otherwise.[3] Renowned theoretical evolutionist Dan Graur, who loudly critiqued the ENCODE project results, increased his estimate of the human genome functionality to 10-25%.[4] Graur is famous for saying, "If ENCODE is right, then evolution is wrong."[5]

However, just as the Bible says in Psalm 9:15, "In the net which they hid, their own foot is caught," so it happened to the theoretical evolutionists. Global data among diverse people groups for DNA sequence variability across the human genome were inputted into a statistical model of neutral evolution. It was discovered that, at most, only 5% of the human genome could randomly evolve and not be subject to the alleged forces of selection. Fanny Pouyet, the lead author of the published study, stated, "What we find is that less than 5% of the human genome can actually be considered as 'neutral.'"[6] So much for human evolution!

This study is just one more example in a long line of failures where the theoretical models of evolution completely collapsed in light of real-world data. And in this case, the failure was even more spectacular because the statistical model used was based on theoretical evolutionary assumptions. Once again, science confirms the special creation of humanity, just like Genesis describes.

## References

1. Pouyet, F. et al. 2018. Background selection and biased gene conversion affect more than 95% of the human genome and bias demographic inferences. *eLife*. 7: e36317.
2. Tomkins, J. P. ENCODE Reveals Incredible Genome Complexity and Function. *Creation Science Update*. Posted on ICR.org September 24, 2012, accessed October 15, 2018.
3. Rands, C. M. et al. 2014. 8.2% of the Human Genome Is Constrained: Variation in Rates of Turnover across Functional Element Classes in the Human Lineage. *PLOS Genetics*. 10 (7): e1004525.
4. Graur, D. 2017. An Upper Limit on the Functional Fraction of the Human Genome. *Genome Biology and Evolution*. 9 (7): 1880-1885.
5. Klinghoffer, D. Dan Graur, Darwin's Reactionary. *Evolution News & Science Today*. Posted on evolutionnews.org June 21, 2017, accessed October 15, 2018.
6. A genome under influence: The faulty yardstick in genomics studies and how to cope with it. Swiss Institute of Bioinformatics press release. Posted on sib.swiss October 9, 2018.

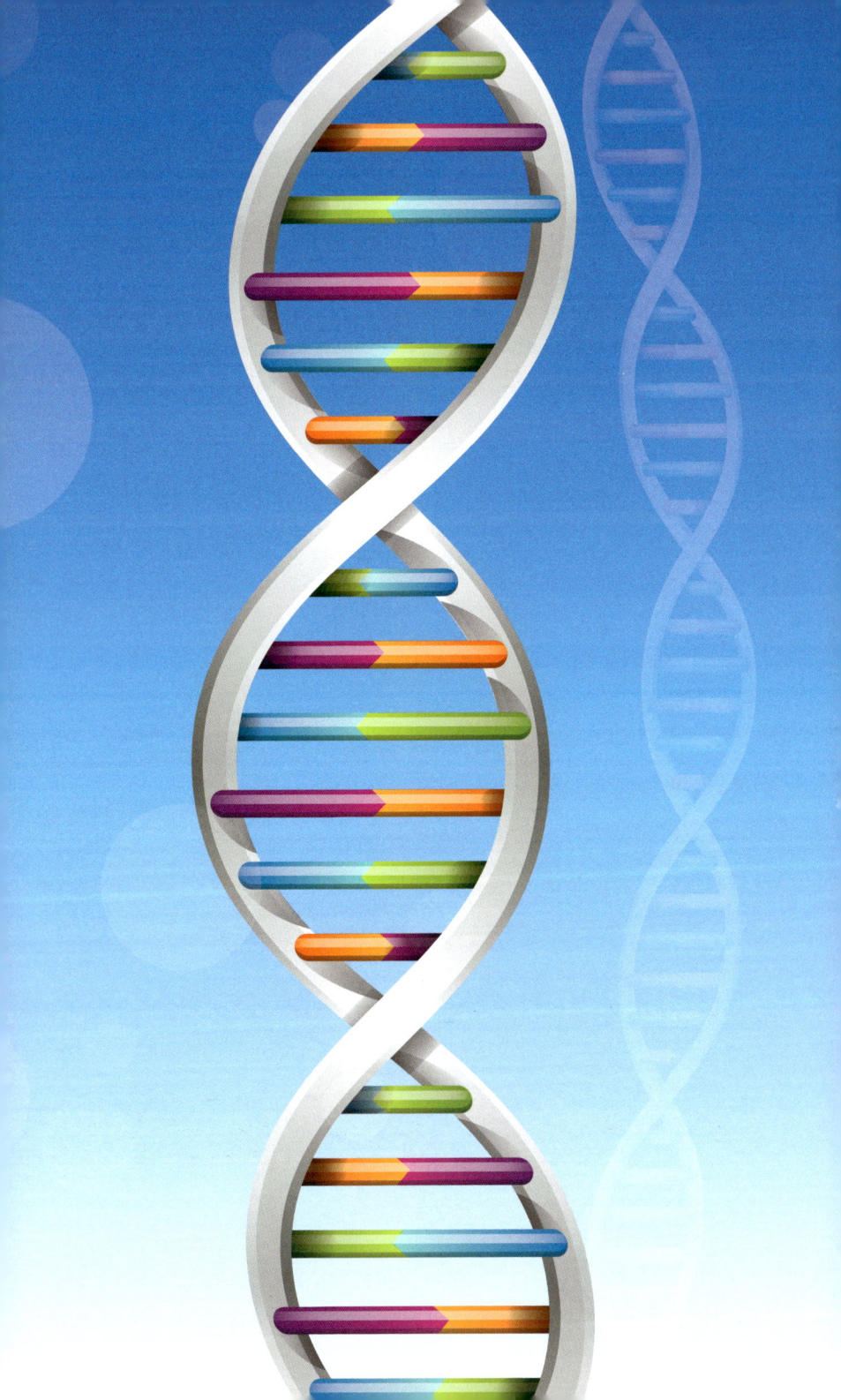

# 5

# SCIENTISTS DISCOVER ANOTHER FUNCTION OF "JUNK DNA"

Frank Sherwin, D.Sc. (Hon.)

- Due to preconceived notions about evolution, many geneticists believed that much of the human genome was nonfunctional.
- But this concept was not based on observation or testing.
- As the science of genetics progresses, scientists continue to discover complex function for areas of the genome that were once considered leftover evolutionary junk.
- Telomerase gene research is just the latest example. Creation scientists predict many more in the future.

For decades, evolutionists suggested that huge sections of our genome did not actively code for proteins or polypeptides—and so were useless, or "junk DNA." But further research uncovered functions for these "worthless" DNA sequences.[1] One creation scientist stated:

> To conclude that a DNA sequence has no function (i.e., that it is "flotsam and jetsam" or "junk"), a scientist must have tested every

base pair (the four DNA base pairs are A, T, G, and C) in the human genome (the totality of our DNA sequence) for function. This is an impossible task.[2]

Jiyue Zhu and his team published an important paper on junk DNA.[3] Zhu is an authority on the process of aging and said:

> These DNA sequences tend to be considered as "junk DNA" or dark matters in our genome, and they are difficult to study. Our study describes that one of those units actually has a function in that it enhances the activity of the telomerase gene.[4]

Telomeres are large parts of DNA that average anywhere between 5,000 to 15,000 bases long in the human genome.[5,6] (Bases are like the letters of the genetic code.)

> The telomerase gene controls the activity of the telomerase enzyme, which helps produce telomeres, the caps at the end of each strand of DNA that protect the chromosomes within our cells. In normal cells, the length of telomeres gets a little bit shorter every time cells duplicate their DNA before they divide. When telomeres get too short, cells can no longer reproduce, causing them to age and die.[4]

Zhu and his team were studying a DNA region deemed "junk" called VNTR2-1. Its function seems to drive the activity of the telomerase gene. This sequence prevents aging in certain types of cells and might help us understand how cancer occurs.

Because of this, scientists believed that if a person had long telomeres, then their aging slowed and they would live longer. But this belief is overly simplistic.

> Zhu said it was worth noting that having a shorter [telomere] sequence does not necessarily mean your lifespan will be shorter, because it means the telomerase gene is less active and your telomere length may be shorter, which could make you less likely to develop cancer.[4]

Further research is pending. Regardless, as predicted by creationists, "the false evolutionary paradigm of 'junk DNA' has been debunked in favor of a new model, one containing pervasive functionality and network complexity. The reality of this seemingly unending complexity is just beginning to be revealed—and it points directly to an omnipotent Creator."[7]

### References

1. Sherwin, F. 2007. Revealing Purpose in "Junk" DNA. *Acts & Facts*. 36 (8): 13.
2. Jeanson, N. 2013. Does "Junk DNA" Exist? *Acts & Facts*. 42 (4): 20.
3. Xu, T. et. al. 2021. Polymorphic tandem DNA repeats activate the human telomerase reverse transcriptase gene. *Proceedings of the National Academy of Sciences*. 118 (26): e2019043118.
4. The potential role of 'junk DNA' sequence in aging, cancer. MedicalXpress. Posted on medicalxpress.com July 23, 2021, accessed August 12, 2021.
5. Tomkins, J. 2012. Internal Telomere Sequences: Accidents of Evolution or Features of Functional Design? *Acts & Facts*. 41 (2): 6.
6. Tomkins, J. 2016. Internal Telomere-like Sequences Are Abundant and Functional. *Acts & Facts*. 45 (10): 9.
7. Tomkins, J. Human Genome 20th Anniversary—Junk DNA Hits the Trash. *Creation Science Update*. Posted on ICR.org April 12, 2021, accessed August 10, 2021.

*Rhesus macaques*

# 6

# APE AND HUMAN SPIT ARE RADICALLY DIFFERENT

Jeffrey P. Tomkins, Ph.D.

In their quest to find evolutionary similarities between humans and apes, scientists compared DNA, proteins, anatomy, behavior, and every other conceivable feature. But many of these attempts showed a huge chasm of dissimilarity with no distinct evolutionary connection. One study comparing saliva between humans and apes shows the uniqueness of humans and the failure of evolutionary reasoning.[1]

Your saliva is highly designed. It is a precise combination of important proteins required to preprocess food in your mouth before entering your digestive tract. Human saliva also contains other specific proteins needed for the maintenance of tooth mineralization and protection from microbial pathogens. The total of the complement of proteins in saliva is called the salivary proteome.

In one research study, scientists compared the salivary proteomes of humans with two ape species considered to be our closest living evolutionary relatives: chimpanzees and gorillas. They also included monkeys (rhesus macaque) as an evolutionary outgroup—an alleged distant relative.

Researchers noted that the first major difference was that human saliva is more watery and diluted than ape saliva, and the overall concentration of proteins is lower. Human saliva contains half the amount of proteins compared to apes and macaques.

The next thing the scientists observed was that the concentrations of the major groups of proteins are markedly different between humans and apes. The researchers also noted that human-specific proteins were found that do not exist in apes. Overall, the salivary proteomes were distinctly different, not only between humans and apes, but also between chimps, gorillas, and macaques.

The researchers concluded, "We discovered unique protein profiles in saliva of humans that were distinct from those of nonhuman primates." They also said, "Certain properties and components of human and nonhuman primate saliva might have evolved in a lineage-specific manner."[1] The term "lineage specific" means there was no evolutionary overlap; each human, ape, and monkey salivary proteome is unique.

This observation doesn't line up with evolution but fits well with Genesis, which tells us that the Lord Jesus created each type of creature after its kind.[2] Humans, chimps, gorillas, and macaques are unique kinds, and both science and Scripture continue to confirm this biological truth.

*References*

1. Thamadilok, S. et al. 2020. Human and Nonhuman Primate Lineage-Specific Footprints in the Salivary Proteome. *Molecular Biology and Evolution*. 37 (2): 395-405.
2. Genesis 1:21, 25

*Australopithecus africanus*

# 7

## *AUSTRALOPITHECUS* ATE LIKE AN APE

Timothy Clarey, Ph.D., and Jeffrey P. Tomkins, Ph.D.

Conventional scientists continually try to humanize ape fossils to bridge the wide gap between humans and apes to fit their worldview. But research published in *Nature* reconfirms that apes have always been apes. The missing links remain missing.

Many scientists claim *Australopithecus africanus* lived about two or three million years ago in open grassland and forests of South Africa.[1] *A. africanus* is one of several *Australopithecus* species that are similar to chimpanzees in their anatomy. Evolutionists use this similarity to support human evolution.

Conventional scientists routinely confirm *Australopithecus* species as nothing more than apes, and this *Nature* study is no different.[2,3] Dr. Renaud Joannes-Boyau and colleagues reported that *A. africanus* exhibited seasonal eating patterns similar to modern apes.[1] The authors noted:

> Although seasonal changes in ecosystems dominated by tropical grassland (frequently referred to as the savannah biome) are associated with only variations in temperature, important oscillations in rainfall produce lengthy dry and wet periods. This has a con-

35

siderable effect on food availability, and leads to long alternating periods of abundance and scarcity of nutritious food.[1]

To examine *A. africanus* fossils for seasonal food variations, the scientists conducted elemental mapping on two first molars from one *A. africanus* specimen (StS 28) and on a canine tooth from a second specimen (StS 51). Their methods included detailed measurements of barium, calcium, lithium, and strontium ratios. This allowed them to create a comprehensive record of variations in food intake during tooth development.[1] For example, barium concentration was heavily tied to the volume of breast milk intake while the ape was growing.

Their results showed overlapping and highly cyclical patterns in Ba/Ca, in Sr/Ca, and in Li/Ca ratios, indicating repeated behavior over time.They further noted, "A similar recurring pattern in Li/Ca, Ba/Ca and Sr/Ca ratios has previously been observed in modern wild orangutans (both *Pongo abelii* and *Pongo pygmaeus*) up to nine years of age."[1]

The authors interpreted these patterns as reflective of seasonal (wet and dry seasons) food adaptations in a grassland environment and fluctuations in food availability. But most importantly, the authors confirm the ape-like nature of australopiths in contrast to humans:

> Similarly, the Li/Ca banding pattern—which is also found in modern orangutans and (to a lesser extent) baboons, but is rarely seen in modern *Homo* [human] samples or in the

non-primate mammals that were analysed—suggests complex physiological adaptations to cyclical periods of abundance and starvation.[1]

*Australopithecus africanus* ate like an ape because it was an ape. Once again, no evidence shows that *A. africanus* was related to humans. Honest evolutionists themselves admitted this fact in a 2016 Royal Society research paper titled "From *Australopithecus* to *Homo*: the transition that wasn't." In this paper, the authors stated:

> Although the transition from *Australopithecus* to *Homo* is usually thought of as a momentous transformation, the fossil record bearing on the origin and earliest evolution of *Homo* is virtually undocumented.[4]

The Lord Jesus made apes, then made Adam and Eve, who were fully human from the start.

*References*

1. Joannes-Boyau, R. et al. 2019. Elemental signatures of *Australopithecus africanus* teeth reveal seasonal dietary stress. *Nature*. 572: 112-115.
2. Tomkins, J. P. *Australopithecus* Was a Well-Adapted Tree Climber. *Creation Science Update*. Posted on ICR.org November 12, 2012, accessed August 7, 2019.
3. Tomkins, J. P. Australopith Child Gets an Academic Spanking. *Creation Science Update*. Posted on ICR.org September 24, 2014, accessed August 7, 2019.
4. Kimbel, W. H. and B. Villmoare. 2016. From *Australopithecus* to *Homo*: the transition that wasn't. *Philosophical Transactions of the Royal Society B*. 371 (1698): 20150248.

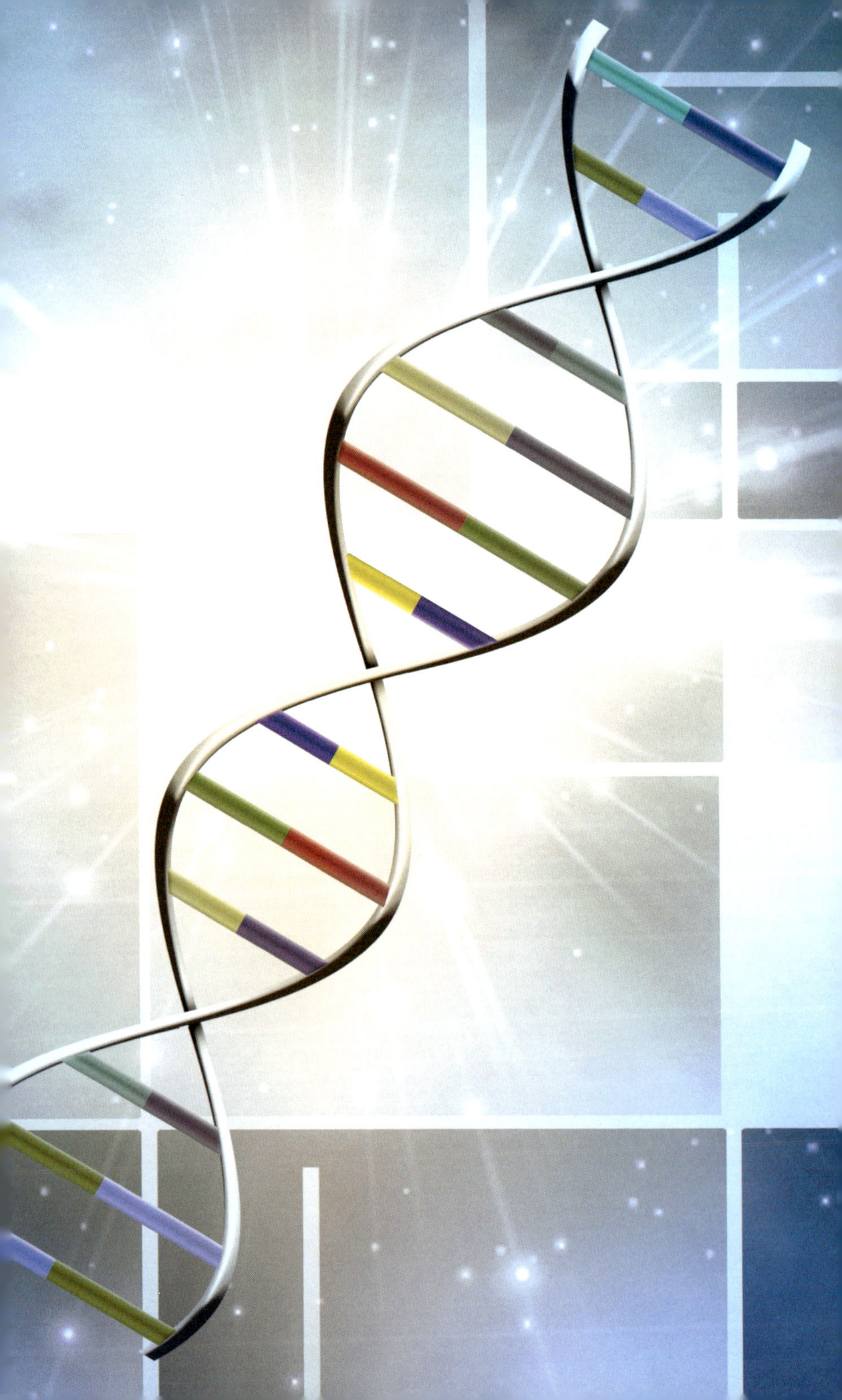

# 8

## CODONS ARE FUNCTIONAL AFTER ALL

Jeffrey P. Tomkins, Ph.D.

- Evolutionists have long thought that DNA can mutate and create new traits.
- They speculated that since codons are variable, they can develop new traits through random chance.
- Research indicates that codons contain multiple layers of valuable and functionally specific genetic information.
- This profound genetic language can only come from a Master Designer.

A popular meme of evolutionary speculation is that many DNA sequences in the genome can freely mutate and create new selectable traits that help creatures evolve. Evolutionists initially applied this belief to protein-coding regions in genes.

Proteins are chains of amino acids. Chromosomes contain regions of code that define how proteins are built. Copies of genes are made using RNA and then processed to contain only the coding regions. These RNA messages are taken out of the cell's nucleus (which houses the chromosomes) and transported into the cytoplasm where specialized machines called ribosomes make proteins. When an RNA is decoded, each three-base sequence, called a codon, specifies a single amino acid in the protein sequence.

There are 61 codons compared to only 20 amino acids. The first two RNA bases in the sequence stay the same, but the third base is variable. For example, the codons GGA, GGC, GGU, and GGG specify an amino acid called glycine, even though the third base is different. As a result, the third base was deemed degenerate and called codon wobble. Evolutionists originally believed that this variability in the third base left room for evolution to work its magic since they thought DNA at these "degenerate sites" could mutate without affecting the resulting protein. They believed codons possess redundancy since they greatly outnumber amino acids.

Codon degeneracy has been presented for years as a viable place in the genome where evolution can occur and be measured, but research discoveries increasingly discredited this concept. Perhaps the most exciting discovery is that other codes are embedded within and overlie the codons.

One study found that a different set of code overlying the codons instructs cellular protein machinery called transcription factors (which control the expression of genes) where to latch onto the DNA inside genes.[1] While one group of codons delineates the amino acid order in a protein, the same sequence of DNA letters can also instruct cellular machinery where to bind to the gene to make the RNA copies needed to make a protein. Researchers called these codes duons.

Shortly after the discovery of duons, scientists discovered another set of codons that control the rate of protein manufacturing at the ribosomal machinery. Altering the rate of protein manufacturing plays an important role in properly folding a protein during production.[2]

In addition to two overlying codes in the same sequence, another discovery showed that the third base in codons regulates the rate of RNA being copied from a gene *and* the levels of RNA copies that are made.[3] This has a downstream effect on the amount of protein that is produced.

An additional research report shows that a fourth code exists in the third base of codons that is related to the overall efficiency of the cells' protein production. Since many proteins from many genes are made at once, the fundamental resources allocated to each type of protein (transfer RNAs) are critical. Like factories that make multiple products, all the assembly lines need a steady supply of parts. The processes need to be perfectly orchestrated. This complex coordination and resource distribution are affected by the third base in codons.[4]

Expert human computer programmers can only write a line of code with a single directive. An all-powerful Creator is the only explanation for genetic code with up to four different layers of instruction in the same sequence of information.

*References*

1. Tomkins, J. P. Duons: Parallel Gene Code Defies Evolution. *Creation Science Update*. Posted on ICR.org January 6, 2014, accessed May 9, 2018.
2. Tomkins, J. P. Dual-Gene Codes Defy Evolution...Again. *Creation Science Update*. Posted on ICR.org September 12, 2014, accessed May 9, 2018.
3. Tomkins, J. P. Codon Degeneracy Discredited Again. *Creation Science Update*. Posted on ICR.org October 13, 2016, accessed May 9, 2018.
4. Frumkin, I. et al. 2018. Codon usage of highly expressed genes affects proteome-wide translation efficiency. *Proceedings of the National Academy of Sciences.* 115 (21): E4940-E4949.

# 9

# DENISOVAN EPIGENETICS REVEALS HUMAN ANATOMY

Jeffrey P. Tomkins, Ph.D.

Using genetic data, one study sought to reconstruct the facial features and anatomy of a group of humans known as the Denisovans. In the evolutionist's own words who did the study, "Denisovans are an extinct group of humans."[1] And the research shows exactly that.

Denisovan fossils are represented by only a few teeth, a finger bone, a bit of a mandible (jawbone), and either a leg or an arm bone fragment. These isolated bits and pieces were found in two locations. One was a Russian cave in the Siberian Altai Mountains close to the borders of Kazakhstan, China, and Mongolia. The other location was farther south in a cave on the Tibetan Plateau. From these teeth and bone fragments, DNA was sequenced and compared to modern human groups.

Denisovan DNA is distinctly human and was found to be closely related to people groups across Asia, including Southeast Asian islands. Denisovan DNA, like Neanderthal DNA, is now considered another variant of the human genome. In fact, evolutionists often acknowledge that anatomically modern

*Denisovan mandible fossil*

humans, Neanderthals, and Denisovans interbred with each other.

Neanderthals are known for having large brains, prominent brow ridges, and sloping foreheads (traits still found among modern humans).[2] However, little is known about what Denisovans looked like because of the limited fossil material. Scientists attempted to reconstruct what a Denisovan might have looked like, using a combination of epigenetics and genetics. It's no surprise that the reconstruction looks fully human.[1] The researchers claimed that the reconstructed female Denisovan had an elongated face and a wide pelvis like Neanderthals, but also had a laterally wider head.

As is typical with evolutionists, they take information from a single individual and then make broad statements to entire populations. Taking DNA data and then inferring what someone looks like from epigenetic modifications has not been shown to be a valid methodology. However, one thing we do know for sure: This is just one more study showing that humans have always been humans.

*References*

1. Gokhman, D. et al. 2019. Reconstructing Denisovan Anatomy Using DNA Methylation Maps. *Cell Press*. 179 (1): 180-192.
2. Tomkins, J. P. 2019. Recent Humans with Archaic Features Upend Evolution. *Acts & Facts*. 48 (4): 15.

*Denisovan cave, Siberia*

# 10

## DENISOVAN DNA SHOWN TO BE HUMAN...AGAIN

Jeffrey P. Tomkins, Ph.D.

Denisovans are ancient humans represented by various teeth and a finger bone found in a Siberian cave. Their fame is largely based on the DNA extracted from these few fragments of human remains. According to evolutionists, Denisovans are more closely related to Neanderthals than modern humans. But their DNA is essentially human, and people worldwide carry many of the same gene variants found in Denisovans.

Why are Denisovans, Neanderthals, and their DNA constantly in the news, other than that the researchers in this popular academic niche need to keep bringing home a paycheck? The mere presence of these enigmatic ancient humans further complicates the speculative field of "recent" human evolution—giving researchers an ongoing knot to untangle. The fuzzy phrase "recent human evolution" refers to the past million years, give or take a few hundred thousand.

Conventional scientists think Africa was the cradle of human evolution. So, they say that *Homo erectus* (early ancient humans) migrated out of Africa in multiple waves, followed by waves of anatomically modern humans. All were basically human, and their fragmentary skeletal remains are well within the range of variation found in modern humans. In fact, the supposedly oldest ancestor of human migration (*Homo erectus*) has been associated with complex tool construction and the use of fire, and their remains were found in remote oceanic island locations in South Asia only accessible by boat.[1,2] And the alleged periods of existence for *Homo erectus*, according to the evolutionists' own dating methods, overlap with both anatomically modern humans and Neanderthal humans.

Here's the evolutionary story. *Homo erectus* migrated out of Africa about one million years ago. About 500,000 years later, the ancient human ancestor of Neanderthals and Denisovans made the same migration. Then, about 120,000 years ago, anatomically modern humans migrated to Europe and Asia, where they interbred with Neanderthals and Denisovans already living there.

Research confirms the results of previous studies. It shows that Denisovan-based regions of the genome are common in Asian people groups and that they interbred with modern humans.[3] Interestingly, the largest proportion of Denisovan DNA in the human genome was found in people from the remote islands of Papua New Guinea, a long way from the Denisovan cave in Siberia. Amazingly, these island people also had a healthy dose of Neanderthal DNA. No good evolutionary reason for this anomaly was offered.

According to the data, all humans belong to the same genetic group. This agrees perfectly with the biblical view and agrees with other studies that show the human genome can be no more than 5,000 years old.[4,5] After the global Genesis Flood about 4,500 years ago, Noah's three sons and their wives formed the basis for repopulating the planet. Their descendants led to the 70 people groups comprising the Table of Nations.[6] As these peoples dispersed from the Mediterranean region across the world, they took their variants of the human genome with them. Thus, it's not surprising that evolutionists cannot make sense of this genetic data with their convoluted out-of-Africa dispersal model.

*References*

1. Lubenow, M. 2004. *Bones of Contention.* Grand Rapid, MI: Baker Books.
2. Browning, S. R. et al. 2018. Analysis of Human Sequence Data Reveals Two Pulses of Archaic Denisovan Admixture. *Cell.* 173 (1): 53-61.
3. Tomkins, J. Genetics Research Confirms Biblical Timeline. *Creation Science Update.* Posted on ICR.org January 9, 2013, accessed March 19, 2018.
4. Tomkins, J. Human DNA Variation Linked to Biblical Event Timeline. *Creation Science Update.* Posted on ICR.org July 23, 2012, accessed March 19, 2018.
5. Tomkins, J. 2014. Genetic Entropy Points to a Young Creation. *Acts & Facts.* 43 (11): 16.
6. Genesis 10

*Charles Darwin*

# 11

# DNA SCIENCE DISPROVES HUMAN EVOLUTION

Jeffrey P. Tomkins, Ph.D.

The Bible describes humans as being created in the image of God—the pinnacle of His creation. In contrast, conventional scientists have put much effort into claiming a bestial origin for man.

The idea that humans evolved from ape-like creatures was first widely promoted by Jean-Baptiste Lamarck in the early 1800s. Twelve years after *The Origin of Species*, Charles Darwin promoted this same theory in *The Descent of Man*. Thomas Huxley, a friend of Darwin, also did much to popularize this idea. Since then, the conventional scientific community has treated the still-hypothetical idea of human evolution as a fact.[1]

Since Darwin's famous publication, we still have no fossil evidence demonstrating human evolution. Darwin believed such fossils would eventually be found, but that has not been the case. The following quotes from evolutionists accurately express the state of affairs regarding the fossil record and its lack of support for human evolution.

The evolutionary events that led to the origin of the *Homo* lineage are an enduring puzzle in paleoanthropology, chiefly because the fossil record from between 3 million and 2 million years ago is frustratingly sparse, especially in eastern Africa.[2]

But with so little evidence to go on, the origin of our genus has remained as mysterious as ever.[3]

The origin of our own genus remains frustratingly unclear.[4]

**The Evolution of Human-Chimp DNA Research**

Although paleontological evidence has been lacking, evidence supporting human evolution was thought to be found in the DNA of living apes and humans. The popular myth of human-chimpanzee DNA similarity along with ongoing research show an unbridgeable chasm between the human and chimpanzee genomes.

DNA is a double-stranded molecule that, under certain conditions, can be denatured—i.e., "unzipped" to make it single-stranded—and then allowed to zip back up. During the initial stages of DNA science in the early 1970s, scientists used crude and indirect techniques to unzip mixtures of human and chimpanzee DNA. They montored how fast the DNA would zip back up compared to unmixed samples.[5] Based on these studies, they declared that human and chimpanzee DNA were 98.5% similar.

But only the most similar protein-coding regions of the genome (called single-copy DNA) were compared, which is an extremely small portion—less than 3%—of the total genome. Later, an evolutionary colleague discovered that the authors of these studies had manipulated the data to make the chimpanzee DNA appear more similar to human than it really was.[6] These initial studies established a fraudulent gold standard of 98.5% DNA similarity between humans and chimps, but also established the shady practice of cherry-picking only the most similar data. The idea of nearly identical human-chimp DNA similarity was born and used to bolster the myth of human evolution, something that the lack of fossil evidence was unable to accomplish.

As DNA sequencing advanced, scientists compared the actual order of DNA bases (nucleotides) between DNA sequences from different creatures. This was done in a process in which similar DNA segments could be directly matched or aligned. The differences were then calculated.

Little progress was made in comparing large regions of DNA between chimpanzees and humans—until the genomics revolution in the 21st century when scientists developed new technologies to sequence the human genome. Between 2002 and 2005, a variety of reports was published that, on the surface, seemed to support the 98.5% DNA similarity myth.

However, a careful analysis of these publications showed that the researchers were only including data on the most highly aligning sequences and omitting gaps and regions that did not align.[5] Once again, we had the same problem of cherry-picking the data that support evolution while ignoring everything else. However, at least three of these papers described the amount of nonsimilar data that was thrown out. When those missing data were included, overall DNA similarity between humans and chimpanzees (depending on the paper) was only about 81 to 87%!

Determining DNA similarity between humans and chimpanzees isn't a trivial task. One of the main problems is that the current chimpanzee genome wasn't constructed based on its own merits. When DNA is sequenced, it's produced in millions of tiny pieces that must be "stitched" together with powerful computers.

This is difficult to do with large mammalian genomes like the chimpanzee. And very few genetic resources exist to aid the effort compared to those available for the human genome project. Because of this resource issue, a limited budget, and a healthy dose of evolutionary bias, the chimpanzee genome was constructed using the human genome as a guide or scaffold onto which the little DNA sequence snippets were organized and stitched together.[7] Therefore, the current chimpanzee genome appears much more human-like than it really is. One study I wrote showed that individual raw chimpanzee DNA sequences aligned very poorly (if at all) onto the chimpanzee genome that had been assembled using the human genome as a framework.[8] This dramatically illustrates that the constructed chimpanzee genome is not an authentic representation of the actual chimpanzee genome.

Another serious problem with the chimpanzee genome is that it appears to contain significant levels of human DNA contamination. When DNA samples are prepared in the laboratory for sequencing, it's common to have DNA from human lab workers get into the samples. Several conventional studies show that many non-primate DNA sequence databases contain significant levels of human DNA.[9,10]

When I studied these data sets, I discovered that a little over half of the ones used to construct the chimpanzee genome contain significantly higher levels of human DNA than the others.[8] These data sets with apparent high levels of human DNA contamination were the ones used during the first phase of the project that led to the famous 2005 chimpanzee genome publication.[11] The data sets produced *afterward* were added atop the ones in the initial assembly. So, not only was the chimpanzee genome assembled using the human genome as a scaffold, but research indicates that it was constructed with significant levels of contaminating human DNA. This would explain why raw unassembled chimpanzee DNA sequences are difficult to align onto the chimpanzee genome with high accuracy—because it's considerably human-like as a result of contamination.

So, how similar are chimpanzee and human DNA? My research indicates that raw chimpanzee DNA sequences with significantly lower levels of human DNA contamination are on average about 85% identical when aligned onto the human genome. Therefore, based on comprehensive research, chimpanzee and human DNA are no more than 85% similar.

**What Does 85% DNA Similarity Mean?**

So, what does 85% DNA similarity mean? First, it's important to note that a DNA similarity of 99% is required for human evolution to seem plausible. This is based on known current mutation rates in humans and an alleged splitting of humans from a common ancestor with chimpanzees about three to six million years ago. This length of time is a mere second on the evolutionary timescale. Any level of similarity much less than 99% is evolutionarily impossible. This is why evolutionists rely on all sorts of tactics when comparing human and chimpanzee DNA—they must achieve a figure close to 99% or their model collapses.

What if humans and chimpanzees are only about 85% similar in their DNA? Isn't this pretty close, even if it puts evolution out of the picture? In reality, this level of similarity is exactly what one would expect from a creation perspective because of certain basic similarities in overall body plans and cellular physiology between humans and chimpanzees. After all, there's a reason DNA is called *the* genetic code. Just as different software programs on a computer have similar code sections because they perform similar functions, the same similarity exists between different creature genomes. This is not evidence that one creature evolved from another, but rather that both creatures were engineered with similar principles. DNA similarities between different creatures are evidence of common engineered design, and that the differences in these DNA sequences are unexplainable by alleged evolutionary processes is also strong evidence of design.

The Bible says that every living thing was created according to its kind. This fits the clear, observable boundaries we see in nature between types of creatures and the distinct boundaries researchers find in genomes as DNA sequencing science progresses.

Regarding humans, we are a distinctly different kind compared to chimpanzees and other apes, and the one part of creation that stands out above all other living forms because the Bible states, "So God created man in His own image; in the image of God He created him; male and female He created them" (Genesis 1:27).

Evolution is a false paradigm that lacks scientific support and attacks one of the key paradigms of the Bible. Humanity's unique creation in God's image is foundational to why Jesus Christ came to redeem us. Man became corrupt through sin from his original created state—he did not evolve that way from an ape.

*References*

1. Menton, D. 2016. Did Humans Really Evolve from Ape-like Creatures? In *Searching for Adam: Genesis & the Truth About Man's Origins*. T. Mortenson, ed. Green Forest, AR: Master Books, 229-262.

2. Kimbel, W. H. 2013. Palaeoanthropology: Hesitation on hominin history. *Nature*. 497 (7451): 573-574.

3. Wong, K. 2012. First of Our Kind: Could *Australopithecus sediba* Be Our Long Lost Ancestor? *Scientific American*. 306 (4): 30-39.

4. Wood, B. 2011. Did early *Homo* migrate "out of" or "in to" Africa? *Proceedings of the National Academy of Sciences*. 108 (26): 10375-10376.

5. Tomkins, J. and J. Bergman. 2012. Genomic monkey business—estimates of nearly identical human-chimp DNA similarity re-evaluated using omitted data. *Journal of Creation*. 26 (1): 94-100.

6. Marks, J. 2011. The Rise and Fall of DNA Hybridization, ca. 1980-1995, or How I Got Interested in Science Studies. In Workshop on "Mechanisms of Fraud in Biomedical Research," organized by Christine Hauskeller and Helga Satzinger. The Wellcome Trust, London, October 17-18, 2008.

7. Tomkins, J. P. 2011. How Genomes are Sequenced and Why it Matters: Implications for Studies in Comparative Genomics of Humans and Chimpanzees. *Answers Research Journal*. 4: 81-88.

8. Tomkins, J. P. 2016. Analysis of 101 Chimpanzee Trace Read Data Sets: Assessment of Their Overall Similarity to Human and Possible Contamination with Human DNA. *Answers Research Journal*. 9: 294-298.

9. Longo, M. S., M. J. O'Neill, and R. J. O'Neill. 2011. Abundant Human DNA Contamination Identified in Non-Primate Genome Databases. *PLOS ONE*. 6 (2): e16410.

10. Kryukov, K. and T. Imanishi. 2016. Human Contamination in Public Genome Assemblies. *PLOS ONE*. 11 (9): e0162624.

11. The Chimpanzee Sequencing and Analysis Consortium. 2005. Initial sequence of the chimpanzee genome and comparison with the human genome. *Nature*. 437 (7055): 69-87.

# 12

## DOES RECENT RESEARCH SUPPORT HUMAN EVOLUTION?

Frank Sherwin, D.Sc. (Hon.)

- Scientific research continues to fail in uncovering evidence for human evolution.
- Humans and chimps are about 10 times more genetically different than evolutionists usually claim—15% rather than 1.5%.
- Out-of-place fossils keep disrupting supposed evolutionary lines of descent.
- Rather than supporting Darwinian evolution, recent research throws serious doubt on the theory.

In 1997, the Institute for Creation Research ran an *Acts & Facts* article on the lack of compelling evidence regarding our supposed evolution from ape-like ancestors.[1] Years have passed, and it's time to see how the case for human evolution has fared since then.

Not so well. Conventional publications are refreshingly blunt. A 2017 *New Scientist* report stated, "The past 15 years have called into question every assumption about who we are and where we came from. Turns out, our evolution is more baffling than we thought."[2] Another article admitted, "Because fossils are so scarce, researchers do not know what the last common ancestors of living apes and humans looked like or where they originated."[3]

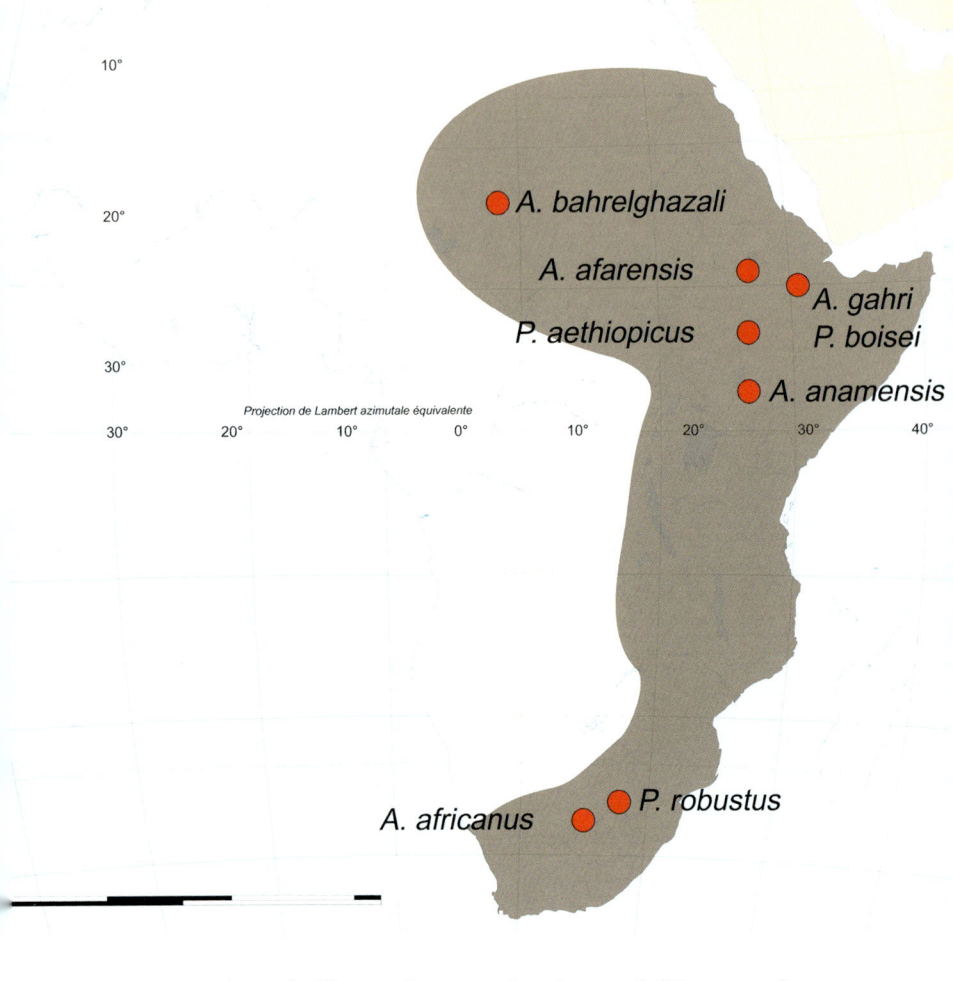

The work of ICR geneticist Dr. Jeffrey Tomkins shows that chimpanzee and human genomes are nowhere near over 98% similar, as traditionally touted by Darwinists.[4] The most recent data show that the human and chimpanzee genomes are no more than about 85% similar.

Even studies comparing the saliva of humans and apes fail to make an evolutionary connection. According to a conventional publication, "We discovered unique protein profiles in saliva of humans that were distinct from those of nonhuman primates."[5]

Anagenesis is an evolutionary term meaning "speciation," a process in which numerous species originate along a single line of descent. Since 1996 (and well before), the "ancestor-descendant sequence" of human evolution has been plagued with the problem of the child being born before the parent or even grandparent. Satisfying progressions from ape-like creatures to more human-like beings were—and are—constantly challenged by out-of-place fossil discoveries. Examples abound, such as two species of *Australopithecus.*

> We further demonstrate that *A. anamensis* and *Australopithecus afarensis* differ more than previously recognized and that these two species overlapped for at least 100,000 years—contradicting the widely accepted hypothesis of anagenesis....Most importantly, MRD [a newly discovered cranium] shows that despite the widely accepted hypothesis of anagenesis *A. afarensis* did not appear as a result of phyletic [evolutionary] transformation.[6]

In 2015, evolutionists introduced a fossil named *Homo naledi* that was immediately embraced as a human ancestor. Being comfortable with a date of three million years, "Prof [Lee] Berger [said] *naledi* could

**DH1**
*H. naledi*

**LES1**
*H. naledi*

be thought of as a 'bridge' between more primitive bipedal primates and humans."[7] But the fossil was shown to be much younger than previously thought, overlapping with anatomically modern humans.[8]

And there's more bad news. In 2002, *Sahelanthropus tchadensis* was discovered and declared an early human relative. But according to *New Scientist*, it "may not have been a hominin at all, but rather was more closely related to other apes like chimps."[9]

When we turn to more recent human ancestors, we see they are 100% human. Biochemist Michael Denton stated:

> Neanderthals and Denisovans must be classed as subspecies or races of *Homo sapiens*, and this would suggest that they may also have had language and relatively high intelligence.[10]

Language and relatively high intelligence? Most of your neighbors would fit that description.

Evolutionary naturalists keep attempting to connect people with chimpanzees and shattered hominid fossils. Regardless, the message of the Bible is clear—

tionists see no anatomical link between *A. anamensis* and mankind. One zoologist admitted, "They are just apes."[4]

Bible-respecting scientists see both australopith varieties as members of the same created ape kind that are now extinct. From this perspective, the ape kinds would have lived simultaneously but perhaps in different places, much like the two chimp varieties on Earth today. The common chimp *Pan troglodytes* has a wide habitat range, but the rarer pygmy chimp (bonobo) *Pan paniscus* appears to only live south of the Congo River in the Congo Basin.

The story of supposedly ancestral *A. anamensis* evolving into *A. afarensis* was easier to believe when *A. anamensis* fossils occurred in rock layers below those of *A. afarensis*. But this newly described *A. anamensis* skull received an evolutionary age that overlaps its supposed descendant by 100,000 years. Instead of living at different times, these two species lived simultaneously. Lead author of the *Nature* paper Haile-Selassie told the Max Planck Institute, "This is a game chang-

*Pygmy chimp*

# 13

## FOSSIL APE SKULL IS A GAME ENDER
Brian Thomas, Ph.D.

A poll of college-age Americans showed that the single most convincing science-based argument for evolution is the lineup of supposed ape-like evolutionary ancestors of mankind.[1] But researchers disagree with each other over the relevance and position of every proposed human ancestor, undermining confidence in this fluctuating and fragmented fossil lineup. Creation researcher Marvin Lubenow called it "the fake parade" in his book *Bones of Contention*.[2] A 2019 ape-fossil study adds more reason to decry this fossil parade as a fake.

A team of experts described a fossil skull that puts a new face and a contradictory age onto an extinct ape variety. This variety was previously known only from teeth and bone fragments.[3] The team discovered the fossil in 2016 in the Afar region of Ethiopia. Early on, they recognized it as an australopithecine ape, but after analysis, they assigned it to the species *anamensis*.

*Australopithecus anamensis* supposedly evolved into *Australopithecus afarensis*—the most famous example of which is nicknamed Lucy—which some evolutionists claim became humans. But many evolu-

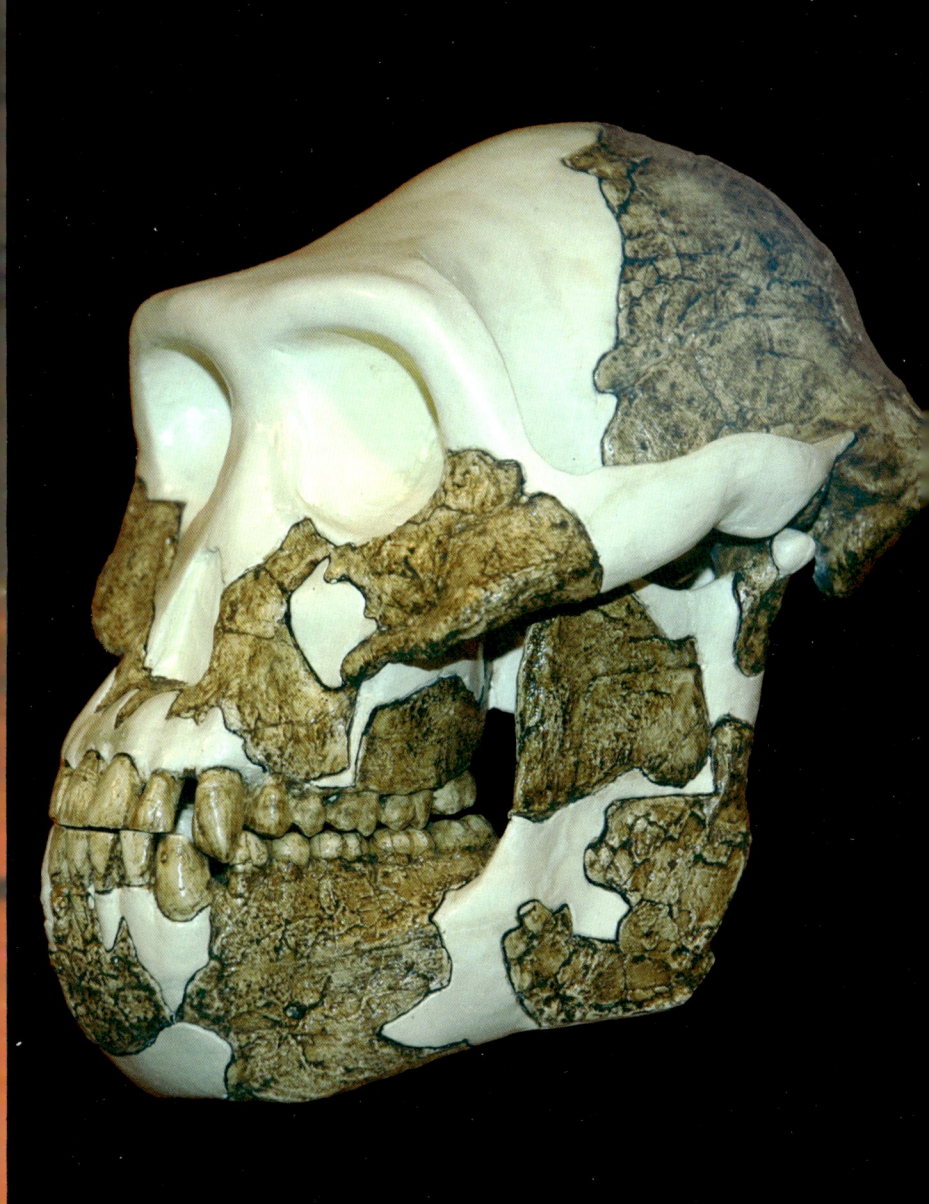

*Australopithecus anamensis*

mankind has been created in God's image since the beginning.

*References*

1. Sherwin, F. 1997. "Human Evolution" An Update. *Acts & Facts.* 26 (9).
2. Barras, C. Who are you? How the story of human origins is being rewritten. *New Scientist.* Posted on newscientist.com August 23, 2017, accessed February 5, 2021.
3. New Study Suggests That Last Common Ancestor of Humans and Apes Was Smaller than Thought. American Museum of Natural History press release. Posted on amnh.org October 12, 2017.
4. Tomkins, J. P. 3-D Human Genome Radically Different from Chimp. *Creation Science Update.* Posted on ICR.org January 7, 2021, accessed February 5, 2021; Tomkins, J. P. 2018. Separate Studies Converge on Human-Chimp DNA Dissimilarity. *Acts & Facts.* 47 (11): 9.
5. Thamadilok, S. et al. 2020. Human and Nonhuman Primate Lineage-Specific Footprints in the Salivary Proteome. *Molecular Biology and Evolution.* 37 (2): 395-405; Tomkins, J. P. Ape Spit Radically Different from Human. *Creation Science Update.* Posted on ICR.org November 12, 2019, accessed February 11, 2021.
6. Haile-Selassie, Y. et al. 2019. A 3.8-million-year-old hominin cranium from Woranso-Mille, Ethiopia. *Nature.* 573: 214-219.
7. Ghosh, P. New human-like species discovered in S Africa. *BBC News.* Posted on bbc.com September 10, 2015, accessed February 12, 2021.
8. Rincon, P. Primitive human 'lived much more recently.' *BBC News.* Posted on bbc.com April 25, 2017, accessed February 12, 2021.
9. Marshall, M. Our supposed earliest human relative may have walked on four legs. *New Scientist.* Posted on newscientist.com November 18, 2020, accessed February 12, 2021; Tomkins, J. P. *Sahelanthropus* Femur Likely Makes It a Chimp. *Creation Science Update.* Posted on ICR.org December 10, 2020, accessed February 11, 2021.
10. Denton, M. 2016. *Evolution: Still a Theory in Crisis.* Seattle, WA: Discovery Institute Press, 198.

er in our understanding of human evolution during the Pliocene."[5]

Talk about an overstatement. Rather, this is a game *ender* for one of the many human evolution narratives. The long overlap in evolutionary time erases notions of an ancestor-descendant relationship between these extinct ape varieties, but leaves intact the Genesis-friendly model of variations within kinds.

*References*

1. Biddle, D. A. and J. Bergman. 2017. Strategically dismantling the evolutionary idea strongholds. *Journal of Creation*. 31 (1): 116-119.
2. Lubenow, M. 2004. *Bones of Contention*. Grand Rapids, MI: Baker Books, 167.
3. Haile-Selassie, Y. et al. 2019. A 3.8-million-year-old hominin cranium from Woranso-Mille, Ethiopia. *Nature*. 573: 214-219.
4. Lewin, R. 1987. *Bones of Contention: Controversies in the Search for Human Origins*. Chicago, IL: University of Chicago Press, 164.
5. A face for Lucy's ancestor. Max Planck Institute. Posted on mpg.de August 28, 2019, accessed September 2, 2019.

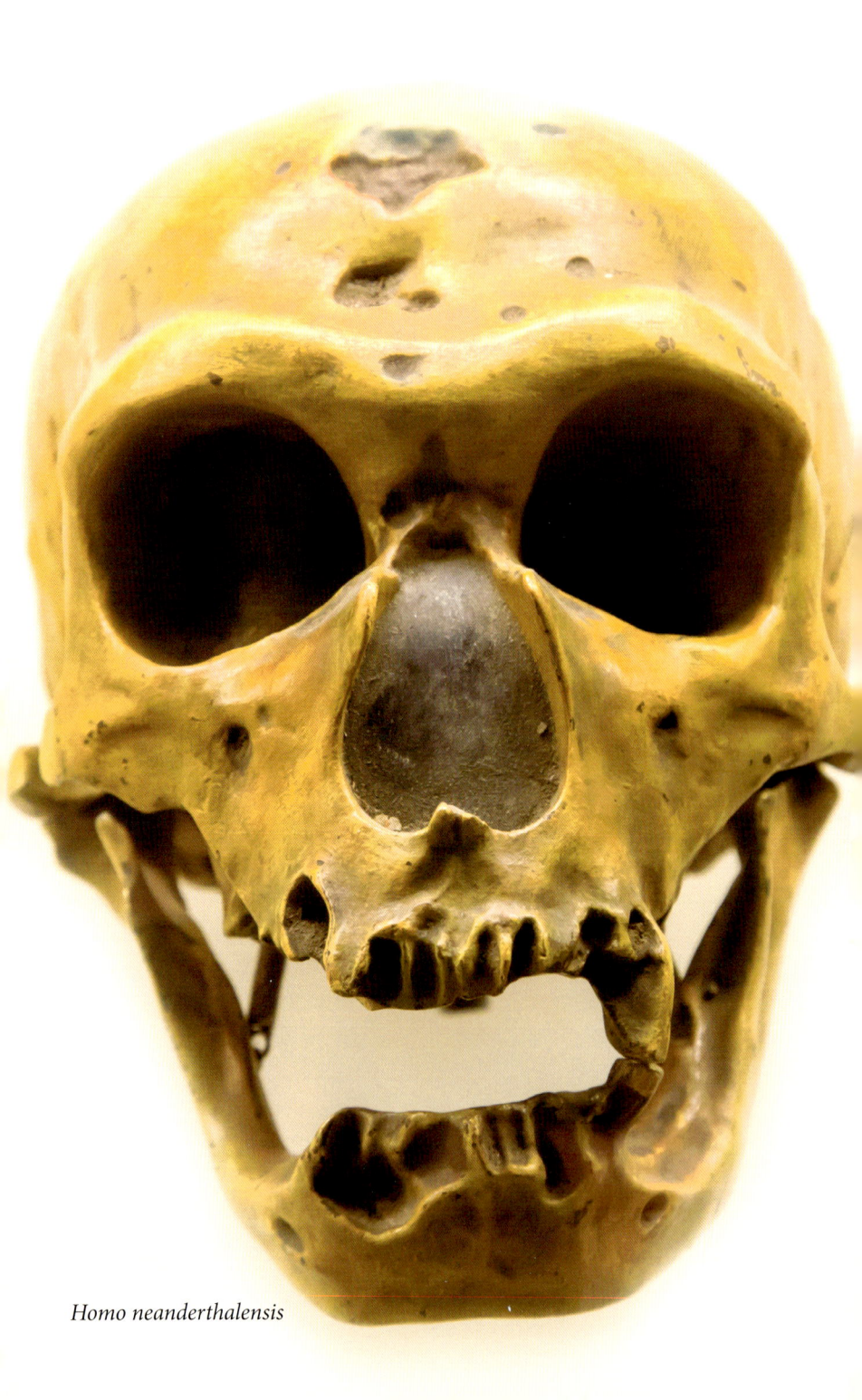

*Homo neanderthalensis*

# 14

# RECENT HUMANS WITH ARCHAIC FEATURES UPEND EVOLUTION

Jeffrey P. Tomkins, Ph.D.

- A "recent" human fossil with "archaic" features found in Mongolia conflicts with the story of evolution.
- Human fossils with archaic features that have recent dates are bad enough for evolutionists, but the fact these traits are found in humans today is far worse.
- So-called archaic humans have always coexisted with modern humans, just as creationists expect.

Ideas shaping the concept of human evolution have largely played out through images. Characters with large brow ridges and sloping foreheads—including *Homo neanderthalensis* and *Homo erectus*—have consistently been depicted as the earliest forms of evolving humans. Now, new fossil evidence is turning the whole paradigm upside down.

A skull fossil found in Mongolia in 2006 was linked to evolutionary icons like *H. neanderthalensis* and *H. erectus* because of its alleged "archaic" features. A 2019 study now dates it at about 34,000 years, which puts it in the same age range (evolutionarily speaking) as very recent humans.[1] This study also extracted mitochondrial DNA from the skull and placed it within the range of modern Eurasian humans. Considering that secular scientists have dated other human skulls with "anatomically modern" features at over 300,000 years,[2] these new findings of "recent" humans with archaic features highlight the abject futility of the human evolution story.

This discrepancy is reminiscent of human skulls found in Kow Swamp, Australia, reported in the journal *Nature* in 1972. In that study, researchers stated, "Analysis of the cranial morphology of more than thirty individuals reveals the survival of *Homo erectus* features in Australia until as recently as 10,000 years ago."[3]

But evolution's problem of human fossils with archaic features persisting into the very recent evolutionary past pales in light of the fact that these traits are still found in living humans. One of the best examples is former Russian boxing champion Nikolai Valuev. A profile picture of Valuev clearly shows he possesses a very prominent brow ridge along with a distinctly sloping forehead.

As things stand, the so-called fossil record for human evolution is still nothing but a collection of apes and humans with no transitional forms linking the two groups. This inconvenient fact was the subject of a 2016 Royal Society research paper bearing the provocative title "From *Australopithecus* to *Homo*: the transition that wasn't."[4]

Numerous studies have shown that australopithecines are extinct apes with many chimp-like anatomical traits. *Homo* is the human genus that includes all of us modern folks along with our assumed archaic ancestors. In the Royal Society paper, the researchers bluntly state:

> Although the transition from *Australopithecus* to *Homo* is usually thought of as a momentous transformation, the fossil record bearing on the origin and earliest evolution of *Homo* is virtually undocumented.[4]

Not only is there no fossil evidence for the evolution of humans from apes, but the so-called archaic features of alleged early evolving humans have in reality coexisted with those of anatomically modern humans throughout the *Homo* fossil record and are even found in humans today. Human skull trait diversity merely demonstrates the created variability that was placed there by the ingenuity of the Creator.

*References*

1. Devièse, T. et al. 2019. Compound-specific radiocarbon dating and mitochondrial DNA analysis of the Pleistocene hominin from Salkhit Mongolia. *Nature Communications.* 10: 274.
2. Hublin, J.-J. et al. 2017. New fossils from Jebel Irhoud, Morocco and the pan-African origin of *Homo sapiens*. *Nature*. 546: 289-292.
3. Thorne, A. G. and P. G. Macumber. 1972. Discoveries of Late Pleistocene Man at Kow Swamp, Australia. *Nature*. 238: 316-319.
4. Kimbel, W. H. and B. Villmoare. 2016. From *Australopithecus* to *Homo*: the transition that wasn't. *Philosophical Transactions of the Royal Society B.* 371 (1698): 20150248.

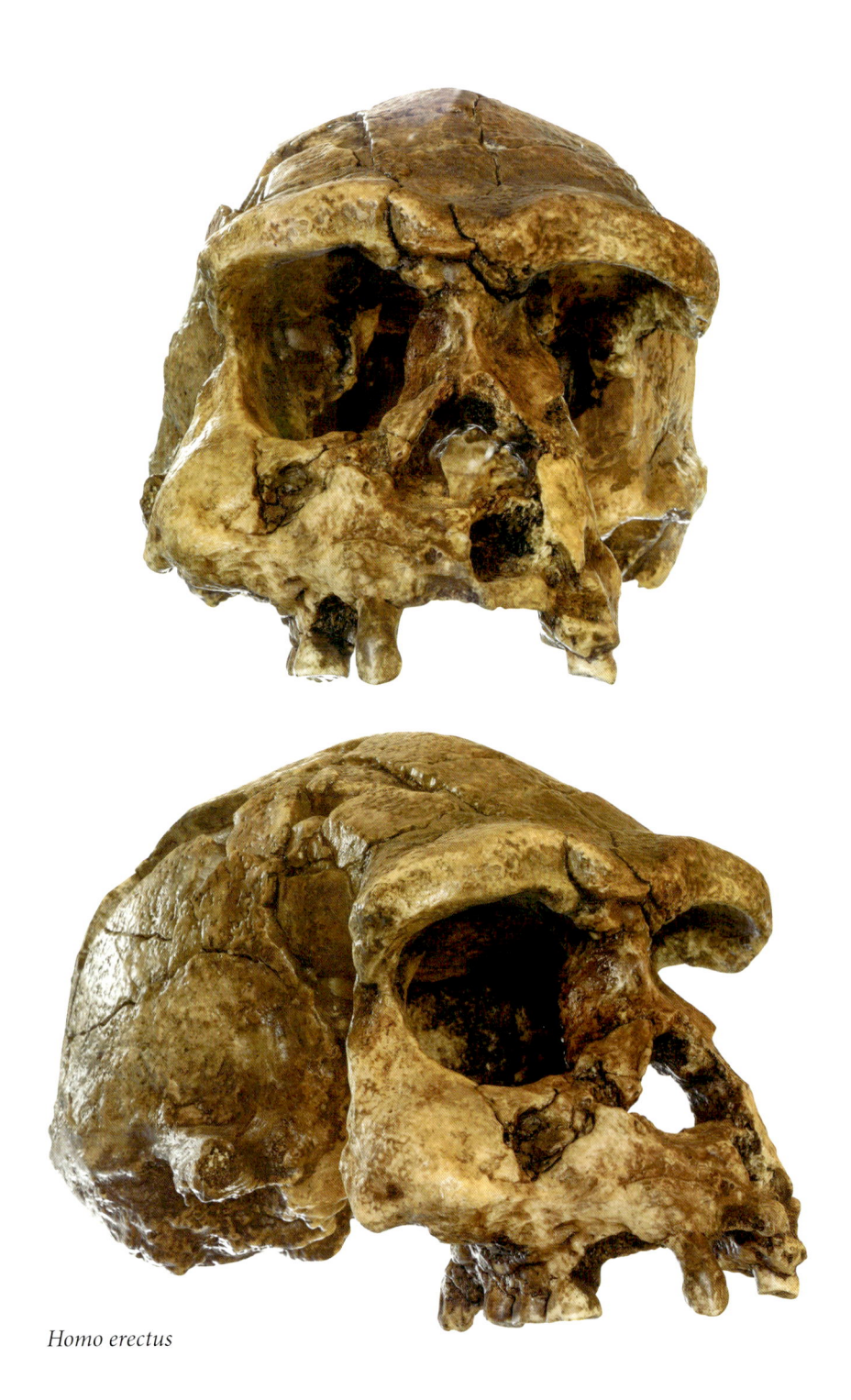

*Homo erectus*

# 15

## *HOMO ERECTUS*:
## THE APE MAN THAT WASN'T

Jeffrey P. Tomkins, Ph.D.

- Many scientists misclassify *Homo erectus* as a transitional species between ape and human.
- So-called "archaic" *Homo erectus* traits like prominent brow ridges and small skulls are found in humans today.
- Conventional dates for some *Homo erectus* discoveries undermine the out-of-Africa model of human migration.
- *Homo erectus* was human, and some of its fossils may have resulted from the Flood.

The archaic human species *Homo erectus* has been portrayed as an important ape-to-man transitional link. However, these fossils don't provide any real evidence of evolution. Many paleontologists and most creationists think their unusual features are nothing more than variants of human traits and not transitional at all. Some fossils have been found in remote island locations far from Africa and dated by conventional calculations at up to 1.9 million years old. This completely derails the evolutionary story that humans migrated out of Africa just a few hundred thousand years ago. A biblical model of human origins fits better with the data.

**How *Homo erectus* Got Going**

The first *Homo erectus* finds were named Java Man and Peking Man.[1,2] Eugène Dubois, a Dutch medical doctor and anatomist, made his famous discovery in 1891 on the island of Java and originally called it *Pithecanthropus erectus*. His Java Man consisted of a skullcap, a thigh bone, and a molar tooth found separately in the same layer of volcanic ash. The skull and thigh bone were about 50 feet apart, but Dubois concluded they belonged to the same individual. An ardent evolutionist and Darwin fan, he immediately claimed he'd found a transitional form. His argument was based primarily on the skullcap's pronounced brow ridge and size. It was smaller than the average modern human but still well within the known variation for humans. The thigh bone was identical to modern humans.

Specimens of Peking Man, also known as *Homo erectus pekinensis*, were discovered between 1923 and 1937 during excavations near Beijing.[1,2] These fossils include six nearly complete crania, 15 partial crania, 11 mandibles (jaws), many teeth, some skeletal bones, and many stone tools.[1,2] They were given evolutionary dates ranging from 680,000 to 780,000 years old.[3,4] Modern human fossils were also found in an upper cave at the same site in 1933.

After these initial discoveries in Asia, similar fossil skulls were found throughout eastern Africa. They were first identified under the name *Homo ergaster*, but now it's widely accepted that *H. ergaster* is the African form of *H. erectus*.

The most complete *H. erectus* fossil was discovered in 1984 near Lake Turkana in Kenya.[5] Known as Turkana Boy, this fossil's skull features were similar to *H. erectus*, but its body was essentially identical to modern humans. Most researchers now agree the skeleton was from a juvenile of about 10 to 12 years of age who would have achieved a normal human height of close to six feet at maturity. This fossil's generally accepted evolutionary age is about 1.6 million years—slightly younger than the Java Man fossils.

The other major group of *H. erectus* fossils was discovered between the Black and Caspian Seas from 1991 to 2005. According to evolutionary dating, the five crania and four mandibles are about 1.8 million years old.[2,6] Although the fossils were placed in the *H. erectus* category, the extreme size and shape variation of the skulls caused controversy. Evolutionists note that if the

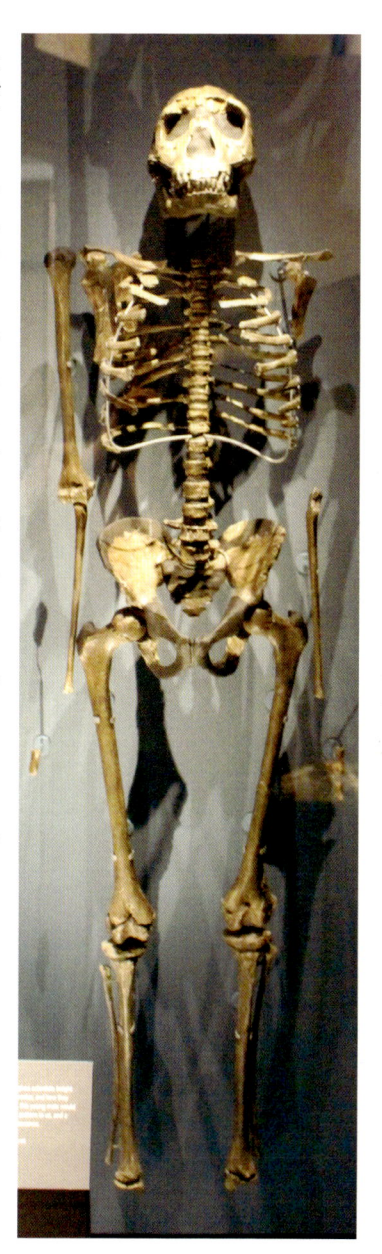

*Turkana Boy*

skulls hadn't been found close to one another and in the same rock layer, they would have been placed in different species categories. Several of the skulls appear to show evidence of disease pathology.

**What Makes a *Homo erectus* a *Homo erectus*?**

The entire story of *H. erectus* is essentially built on about 300 fragmentary fossils. The majority of these are nothing more than partial skulls, teeth, and broken bones. The only nearly complete *H. erectus* fossil is Turkana Boy, whose post-cranial skeleton was found to be almost identical to modern humans.

Based on the diverse skull fragments and a few nearly complete crania, the defining features of *H. erectus* are:

- Prominent brow ridge
- Sloping forehead
- Reduced chin
- More constricted temples than typical humans
- Larger teeth
- Forward-projecting jaw (prognathism)
- Cranial capacities on the lower end of the normal human-size spectrum[2]

*H. erectus* cranial volume is smaller, on average, than modern humans but still within the range. Research has shown that, in general, human or animal intelligence is not based on brain size but on creature-specific organizational properties.[2]

### Recent *Homo erectus* in Australia and China?

In 1972, the fossil remains of about 50 Aboriginal humans were discovered in the Kow Swamp region of Australia.[2,7] The researchers described "archaic" human traits "not seen in recent Aboriginal crania" that closely paralleled the traits of classic *H. erectus*. These included prominent brow ridges, sloping foreheads, prognathism, large teeth, and a minimal chin. But most importantly, along with the reported "archaic" features, the researchers claimed a very recent date (by evolutionary standards) for these fossils. The report stated, "Analysis of the cranial morphology of more than thirty individuals reveals the survival of *Homo erectus* features in Australia until as recently as 10,000 years ago."[7]

In 2006 in Mongolia, researchers found a skullcap whose "analysis shows similarities with Neanderthals, Chinese *Homo erectus*, and West/Far East archaic *Homo sapiens*."[8] Just like the Kow Swamp fossils, the evolutionary dates don't fall within the range of typical *H. erectus*. A later study of the Mongolian fossil gave an age of around 34,000 years old—a time considered very recent in the human

*Kow Swamp*

evolutionary spectrum and on par with the Kow Swamp fossils.

The problem with both of these fossil discoveries is that they have been put in the same age range as recent anatomically modern humans. If they had been dated at one to two million years, they would have been considered *H. erectus* and fit the evolutionary narrative. As things stand, they are considered mere anomalies to be swept under the rug to maintain the evolutionary myth that "archaic" human traits disappeared long ago.

**Archaic Traits Are Still Alive and Well**

Scientists discovered a human skull in Jebel Irhoud, Morocco, defined as having "anatomically modern" features and dated at over 300,000 years old.[9] This, combined with "recent" humans with archaic features (the Kow Swamp and Mongolian fossils), highlights severe inconsistencies in the human evolution story.

If one wants to accept the evolutionary timeline, then *H. erectus* with "archaic" features and humans with anatomically modern features have hung out together on Earth for a long time, even up to the recent past. But it gets worse for the evolutionary picture because "archaic" traits like prominent brow ridges, sloping foreheads, prognathism, and small skulls can still be found in living humans.[10]

**Advanced Culture and Behavior of *Homo erectus***

There is strong anatomical evidence that *H. erectus* was just a variant of the human kind and copious archaeological evidence that its members were highly intelligent and exhibited a broad range of human behaviors. The condensed list below is based on an extensive scientific literature survey published in 2017.[2]

- Watercraft construction and seafaring navigation
- Language and communication skills
- Jewelry manufacture
- Cordage and knot-making
- Manufacture and use of stone and bone tools
- Controlled usage of fire and cooking
- Catching and processing fish
- Development of organized living and occupational spaces
- Art (petroglyphs, figurines, red ochre paint)
- Woodworking
- Coordinated large-game hunting and processing
- Development of clothing from animal skins
- Development of fibers and resins
- Social and family structure
- Care for the elderly and weak

**Homo erectus and the Out-of-Africa Myth**

The *H. erectus* fossil presence in China, southeast Asia, and remote islands like Java causes major problems for the evolutionary migration paradigm. It's obvious that intelligent seafaring humans made the journey over long stretches of open ocean to reach these islands. If we accept the deep-time dating of *H. erectus* up to two million years in remote places like this, then there are severe evolutionary time frame discrepancies.

For one thing, the supposed initial phase of human evolution is represented by the ape-like *Australopithecus*, which overlaps significantly in time with *H. erectus*. If they coexisted, then how could one have been the evolutionary predecessor of the other? Even worse is the problem it presents for the current out-of-Africa model. This model proposes that humans migrated from Africa only about 100,000 to 200,000 years ago. But if that is the case, then how could they have existed on remote southeast Asian islands two million years before that?

***Homo erectus* Was Human After All**

The *H. erectus* fossil record is fragmentary and incomplete, but the bulk of the data indicates this category is simply a variant of the human kind. As mentioned above, "archaic" *H. erectus* traits can still be found in humans today. Even many evolutionists recognize this. One article stated, "If you bumped into a *Homo erectus* in the street you might not recognise them as being very different from you."[11]

So, *H. erectus* was fully human, and the evolutionary narratives and timelines don't make sense even within a secular worldview. How does this fit with the Bible? First of all, the Scriptures clearly teach that God created an ancestral human couple uniquely in His image on the sixth day of the creation week. We also know that death, sin, and corruption entered the picture when Adam and Eve rebelled (Genesis 3). As the human population grew, people became so wicked that God destroyed the world in a global flood. Some of the *H. erectus* fossils might have been humans buried in the highest sedimentary layers of Flood rock, exactly where we would expect them.

Alternatively, some *H. erectus* fossils may have been from specific people groups that dispersed from the Tower of Babel after the Flood. The Kow Swamp burial site would be a good example. When human languages were confused at Babel, the resulting isolated groups would have led to a large number of genetic bottlenecks and lineages of humans with unique suites of trait variations such as skin color, skeletal sizes, and skull shape variations. Some creation scientists speculate that genetic abnormalities would have surfaced more rapidly in small, isolated, inbred populations, which might explain some of the unusual variations seen in *H. erectus* skulls.[2]

Clearly, the biblical account of human creation and Earth history offers a more satisfying framework in which to place human fossil discoveries than the scientifically flawed evolutionary narrative.

*References*

1. Lubenow, M. 2004. *Bones of Contention*. Grand Rapids, MI: Baker Books.
2. Rupe, C. and J. Sanford. 2017. *Contested Bones*. Waterloo, NY: FMS Publications.
3. Shen, G. et al. 2009. Age of Zhoukoudian *Homo erectus* determined with $^{26}Al/^{10}Be$ burial dating. *Nature*. 458: 198-200.
4. Devièse, T. et al. 2019. Compound-specific radiocarbon dating and mitochondrial DNA analysis of the Pleistocene hominin from Salkhit Mongolia. *Nature Communications*. 10: 274.

5. Brown, F. et al. 1985. Early *Homo erectus* skeleton from west Lake Turkana, Kenya. *Nature*. 316 (6031): 788-792.

6. Schwartz, J. H., I. Tattersall, and Z. Chi. 2014. Comment on "A Complete Skull from Dmanisi, Georgia, and the Evolutionary Biology of Early *Homo*." *Science*. 344 (6182): 360.

7. Thorne, A. G. and P. G. Macumber. 1972. Discoveries of Late Pleistocene Man at Kow Swamp, Australia. *Nature*. 238: 316-319.

8. Coppens, Y. et al. 2008. Discovery of an archaic *Homo sapiens* skullcap in Northeast Mongolia. *Comptes Rendus Palevol*. 7 (1): 51-60.

9. Hublin, J.-J. et al. 2017. New fossils from Jebel Irhoud, Morocco and the pan-African origin of *Homo sapiens*. *Nature*. 546: 289-292.

10. Tomkins, J. P. 2019. Recent Humans with Archaic Features Upend Evolution. *Acts & Facts*. 48 (4): 15.

11. Moffat, I. A snapshot of our mysterious ancestor *Homo erectus*. *Phys.Org*. Posted on Phys.org January 29, 2019, accessed August 12, 2019.

# 16

## *HOMO NALEDI*: ANOTHER FAILED EVOLUTIONARY APE MAN

Jeffrey P. Tomkins, Ph.D.

- Lee Berger's 2015 *Homo naledi* fossil discovery contained some 1,550 bone fragments and supposedly documented an intentionally buried new hominid species.

- Even some evolutionists were skeptical of Berger's claims, and bone analysis shows that *Homo naledi* was overall an ape-like creature.

- Dating of the fossils and rocks showed them to be much younger than the two million years the evolutionary story demands.

- This is another failed attempt to demonstrate human evolution.

One of the most confusing and enigmatic "ape-man" finds of the 21st century was *Homo naledi*, discovered by paleoanthropologist Lee Berger. The claims surrounding this discovery have been extolled, criticized, and debated by both evolutionists and creationists. A 2015 article in *The Guardian* highlighted the raging controversy. It was titled "Scientist who found new human species accused of playing fast and loose with the truth."[1]

Since the first journal publication in 2015,[2] much more has been published analyzing the bone fragments and other archaeological and geological aspects of *H. naledi*. As a result, we can take a fresh look at all the data and conclude that conventional scientists used another false ape-man story to support human evolution.

## History of the *Homo naledi* Discovery

The story told by Berger in his book *Almost Human* reveals that a former student mysteriously showed up and convinced him to support an effort to explore caves in the area of South Africa where he was working.[3] The student also persuaded Berger to use the labor of several amateurs experienced in cave exploration. Fortunately for Berger, the amateur explorers could penetrate the lower reaches of the Ris-

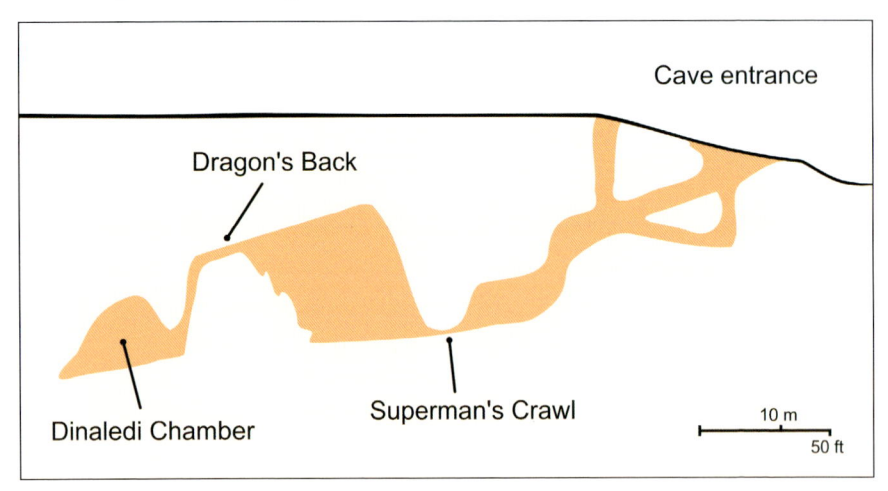

*Figure 1. The Dinaledi Chamber is the lowest room in the Rising Star cave system and can only be accessed through an extremely narrow and nearly vertical chute about 39 feet long.*

ing Star cave system and find a remote chamber littered with fossils. When Berger saw their pictures of the fossils, he thought, "It wasn't human; that much was clear."[3]

As the Rising Star cave system progresses downward, two narrow passages connect the two lowest chambers (Figure 1). When Berger investigated the cave system, he just barely squeezed through the first narrow passage and entered a large chamber called the Dragon's Back. He noticed that the walls were covered with fossils. In his book, he states, "This chamber alone deserved further investigation, but we were here to see fossils farther on."[3]

Numerous fossils were embedded in sediments in the Dragon's Back wall through obvious flooding of the cave system. Berger's initial announcements omitted this highly relevant fact. They claimed the fossils in the chamber below it, the Dinaledi Chamber, had been intentionally buried—not flood-deposited. This chamber contained the fossils that most interested Berger. However, he could not get through the narrow chute to reach them, so he hired a team of six thin, small women to do the fossil excavations.

After several rounds of excavation, the Dinaledi Chamber yielded 1,550 mostly disarticulated bone fragments, plus an undisclosed number of rodent and bird fossils. Berger's team pieced together as much of this hodgepodge of bones as they could and claimed that a total of 15 different individuals were represented. These findings supposedly documented an alleged new hominid species and were published

*Human Origins*

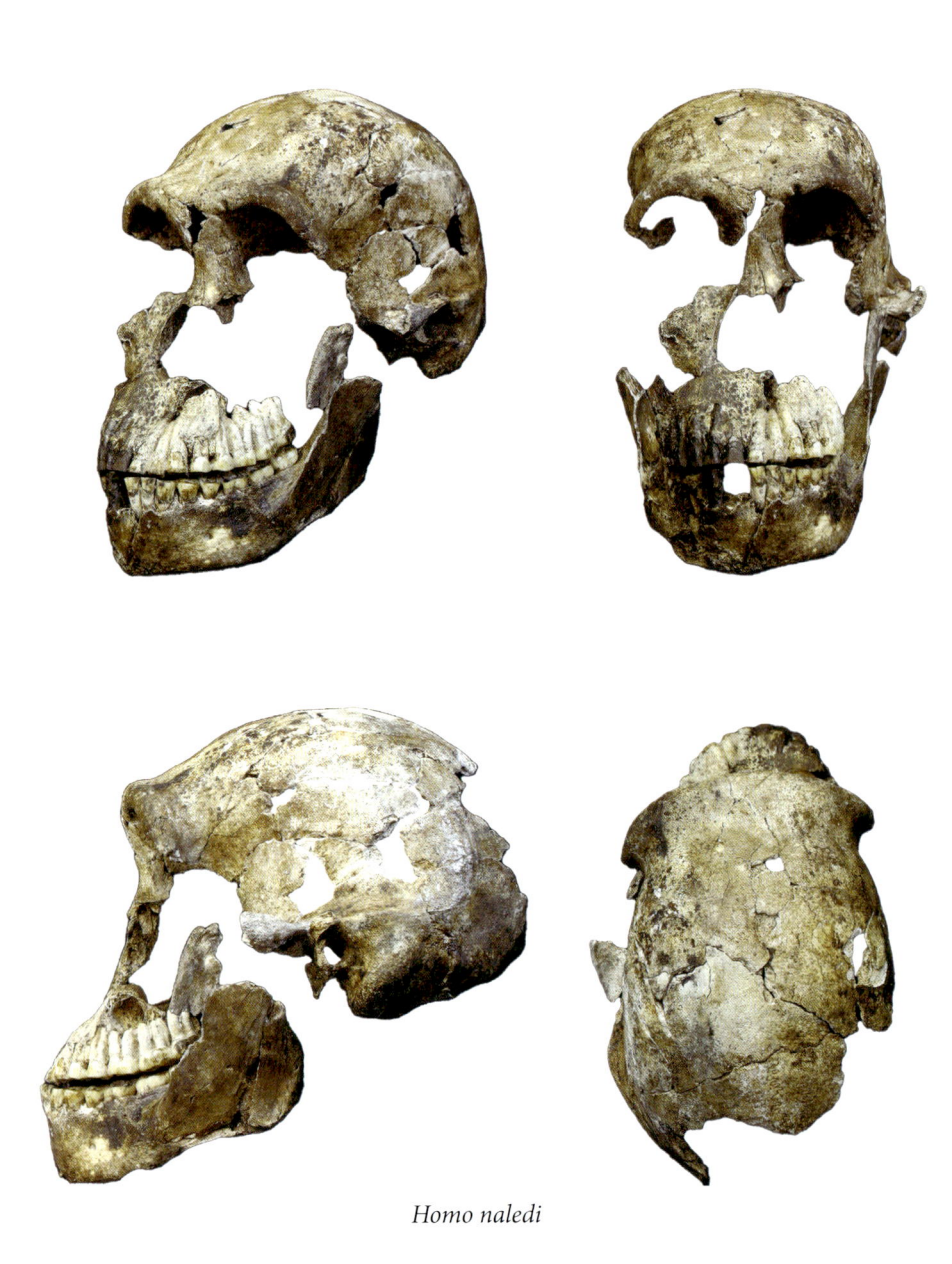

*Homo naledi*

in the lower-tier scientific journal *eLife*.[2] Berger's discoveries and new hominid claims also benefited from popular media coverage provided by *National Geographic* magazine.

However, Berger's discovery soon became controversial. World-famous hominid paleoanthropologist Tim White revealed to the press that the prestigious journal *Nature* had rejected Berger's paper along with its conclusions.[4] In other words, Berger's claims concerning *H. naledi* were met with strong skepticism even among evolutionists.

Another odd twist to the *H. naledi* story is the incriminating revelation made by Berger in his book that his group had known about another section of the cave system containing more hominid fossils that was more easily accessible. However, they kept it quiet while the *H. naledi* story was being formulated. In 2017, Berger's group published a paper detailing the presence of at least three more *H. naledi* fossils in this other section in what is now called the Lesedi Chamber.[5]

**What Was *Homo naledi*?**

Many problems surround the myriad of bone fragments and their reconstruction to supposedly reveal 15 new hominids from the Dinaledi Chamber. We'll examine three. The first problem is homogeneity—whether all the fossils even belong to the same species. Berger and his researchers initially claimed (and still do) that the bones were homogeneous in their representation of a single almost-human species.[2,6]

However, evolutionary biologist Jeffrey Schwartz first noted the extreme nonhomogeneity of the fossils. He believed that the huge mix of bone fragments was too varied to represent a single species. He said, "I could show those images to my students and they would say that they're not the same."[7] Schwartz also claimed that one of the skulls looked like it came from an australopith (ape-like creature), as did certain features of the femurs. In a 2018 paper, Berger and his team state, "The Dinaledi ossicles resemble those of chimpanzees and *Paranthropus robustus* [an ape] more than they do later members of the genus *Homo*."[8]

Since the original 2015 *eLife* publication, numerous research papers have been published, mostly by members of Berger's team. They keep showing that *H. naledi* is nothing more than a suspicious hodgepodge of ape-like bones (*Australopithecus*) and a few human-like bones. These papers analyzed skulls, pelvic remains, leg bones, hands, and feet and give the same confusing story.[6,9-13] One of the few critical papers published outside Berger's group contradicted the claims that *H. naledi* had flat, human-like feet.[14] In addition, a paper analyzing pelvic remains stated:

> Though this species has been attributed to *Homo* based on cranial and lower limb morphology, the morphology of some of the fragmentary pelvic remains recovered align more closely with specimens attributed to the species *Australopithecus afarensis* and *Australopithecus africanus*.[10]

One attempt to bolster *H. naledi* as being almost human involved the study of a skull endocast (a cast of the inside of the cranium). Berger's group claims, "*H. naledi* shared some aspects of human brain organization."[15] They are referring to a human-specific brain region called BA45. Shawn Hurst, one of the study authors, consulted with Dean Falk, a neurobiology specialist in hominid paleontology. Falk disagreed with their conclusions.

> "We agreed on most of the interpretations," she says—but not on the presence of a modern BA45...."I'm not seeing BA45," says Falk. "To me the general shape of the region looks ape-like."[16]

**The Dating Problem**

A second problem concerns the dating of *H. naledi*. When *H. naledi* was first published, there were no official radiometric dates for it—just the speculations of Berger and his team. They stated, "If the fossils prove to be substantially older than 2 million years, *H. naledi* would be the earliest example of our genus that is more than a single isolated fragment."[2] These optimistic speculations were soon dashed against the stones of radiometric techniques.

In 2017, a report was published using six different types of dating techniques.[17] These included radiocarbon (C-14), electron-spin resonance (ESR), uranium-thorium decay (U-Th), and optically-stimulated luminescence (OSL) in a central age statistical model (CAM), and OSL in a minimal age model (MAM). These techniques were applied to bones, teeth, and flowstones in the cave where the fossils were found. Some of the flowstones had partially covered the fossils. Depending on the technique, ages came forth that varied widely from 33,000 to 849,000 years.

The youngest dates were derived from the C-14, U-Th, and ESR dating of the fossil bones and teeth, which gave ages from 33,000 to 146,000 years. The researchers rejected these dates. Instead, they chose the older dates taken from the rocks and the high end of the range from the teeth. The researchers stated:

> By combining the US-ESR maximum age estimate obtained from the teeth, with the U-Th age for the oldest flowstone overlying *Homo naledi* fossils, we have constrained the depositional age of *Homo naledi* to a period between 236 ka and 335 ka.[17]

However, even these cherry-picked dates contradict the evolutionary story. *H. naledi* was supposedly an ancestor of *Homo erectus,* who was an ancestor of modern humans. However, *H. erectus* fossils have been found that supposedly date up to 1.9 million years—long before *H. naledi*.[18] Additionally, evolutionists believe modern humans have been around for the last 300,000 years, making them contemporaries of *H. naledi*.[19] As a result, the researchers of the dating study conceded:

> These age results demonstrate that a morphologically primitive hominin, *Homo naledi*, survived into the later parts of the Pleistocene in Africa, and indicate a much younger age for the *Homo naledi* fossils than have pre-

viously been hypothesized based on their morphology.[17]

## The Intentional Burial Story

A third problem concerns Berger's contention that the bones were intentionally buried. The extremely young dates (by evolutionary standards) pose a severe problem for *H. naledi*. And the story originally put forth by Berger and his team that the bones were intentionally buried has been just as troubling. The companion paper to the original 2015 publication describing the geology at the site stated:

> The fossils are contained in mostly unconsolidated muddy sediment with clear evidence of a mixed taphonomic signature indicative of repeated cycles of reworking and more than one episode of primary deposition.[20]

So, not only were the fossils completely disarticulated and jumbled up in a muddy deposit, they were also mixed with various bird and rodent bones.

As noted earlier, Berger revealed in his book that the walls of the Dragon's Back Chamber above the Dinaledi were covered with unspecified fossils. These were clearly washed in with so much water that they were pushed up and pasted against the sides of the cave. The obvious implication of both the geology and the array of disarticulated creatures in this upper chamber is that a similar flooding event washed all the bones into the lowest chamber.

Even more suspicious is Berger's careful storytelling to support his claim that the *H. naledi* fossils were intentionally buried, but at the same time, he hid the Lesedi Chamber discovery. If his story were true, then the Lesedi Chamber would have been a more logical location for burial since it is more easily accessible and would not require super-gymnastic athletic ability such as that needed to enter the lower Dinaledi Chamber. Also, why are we not told what types of fossils were buried in the Dragon's Back Chamber directly above it? Is it because it contains the same fossil debris as the Dinaledi Chamber below it? This would prove they were all deposited during a cave flooding event.

The muddy, jumbled deposit of bones looks like they were washed in by a local flood, and analysis of the cave's geology shows that it is largely a single deposit.[21] In addition, a machine-learning computer study demonstrated that based on the position of the bones compared to authentic ancient burial sites, *H. naledi* was not intentionally buried.[22] These data also fit well with the fact that no tools or signs of human occupation have been found in the cave, nor are there any signs of using torches to provide light during burial.

Furthermore, a forensic microscopic analysis of the *H. naledi* bones indicates they were fed on by snails that only live in the entrances of caves where light shines.[23] When you combine this with the fact that the smaller *H. naledi* bones were broken up, the real story emerges: these ape-like creatures were likely killed by carnivores and hauled into the entrance of the cave system.[23] Then they were disarticulated during feeding, and their carcasses continued to be scavenged. Eventually, a flood washed the bones, along with those of rodents and birds, into the lowest chamber of the cave.

## Conclusion: Another Failed Attempt at Human Evolution

What can we make of the conflicting conclusions depending on which bone fragments are being evaluated and who is doing the analysis? First, it is likely that most, if not all, of the hominid bones in the Dinaledi and Lesedi Chambers belong to *Australopithecus* (ape-like creatures). It is possible that a small human, perhaps a juvenile, could have been killed by a predator and added to the majority australopith mix. This is entirely feasible given the track record of Lee Berger in the case of his previous *Australopithecus sediba* discovery, which was later determined to likely be a mix of human and mostly ape-like bones.[24]

When you combine the ape-like nature of the fossil bones with the young dates achieved by evolutionary methods, as well as the overwhelming data for carnivory and a cave flooding-based deposition, *H. naledi* stands as another failed attempt at promoting human evolution.

### References

1. McKie, R. Scientist who found new human species accused of playing fast and loose with the truth. *The Guardian.* Posted on theguardian.com October 24, 2015, accessed November 25, 2019.
2. Berger, L. R. et al. 2015. *Homo naledi*, a new species of the genus *Homo* from the Dinaledi Chamber, South Africa. *eLife*. 4: e09560.
3. Berger, L. and J. Hawks. 2017. *Almost Human: The Astonishing Tale of* Homo naledi *and the Discovery That Changed Our Human Story*. Washington, DC: National Geographic.

4. Martin, G. Bones of Contention: Cal Paleo Expert Doubts *Homo Naledi* Is New Species. *California Magazine.* Posted on alumni.berkeley.edu October 1, 2015, accessed November 25, 2019.

5. Hawks, J. et al. 2017. New fossil remains of *Homo naledi* from the Lesedi Chamber, South Africa. *eLife.* 6: e24232.

6. Marchi, D. et al. 2017. The thigh and leg of *Homo naledi. Journal of Human Evolution.* 104: 174-204.

7. Callaway, E. 2015. Crowdsourcing digs up an early human species. *Nature.* 525 (7569): 297-298.

8. Elliott, M. C. et al. 2018. Description and analysis of three *Homo naledi* incudes from the Dinaledi Chamber, Rising Star cave (South Africa). *Journal of Human Evolution.* 122: 146-155.

9. Schroeder, L. et al. 2017. Skull diversity in the *Homo* lineage and the relative position of *Homo naledi. Journal of Human Evolution.* 104: 124-135.

10. VanSickle, C. et al. 2018. *Homo naledi* pelvic remains from the Dinaledi Chamber, South Africa. *Journal Human Evolution.* 125: 122-136.

11. Feuerriegel, E. M. et al. 2017. The upper limb of *Homo naledi. Journal of Human Evolution.* 104: 155-173.

12. Laird, M. F. et al. 2017. The skull of *Homo naledi. Journal of Human Evolution.* 104: 100-123.

13. Williams, S. A. et al. 2017. The vertebrae and ribs of *Homo naledi. Journal of Human Evolution.* 104: 136-154.

14. Li, R. et al. 2019. *Homo naledi* did not have flat foot. *Homo.* 70 (2): 139-146.

15. Holloway, R. L. et al. 2018. Endocast morphology of *Homo naledi* from the Dinaledi Chamber, South Africa. *Proceedings of the National Academy of Sciences.* 115 (22): 5738-5743.

16. Barras, C. Mystery human species *Homo naledi* had tiny but advanced brain. *New Scientist.* Posted on newscientist.com April 24, 2017, accessed November 25, 2019.

17. Dirks, P. H. et al. 2017. The age of *Homo naledi* and associated sediments in the Rising Star Cave, South Africa. *eLife.* 6: e24231.

18. Tomkins, J. 2019. *Homo erectus*: The Ape Man That Wasn't. *Acts & Facts.* 48 (10): 11-13.

19. Richter, D. et al. 2017. The age of the hominin fossils from Jebel Irhoud, Morocco, and the origins of the Middle Stone Age. *Nature.* 546: 293-296.

20. Dirks, P. H. et al. 2015. Geological and taphonomic context for the new hominin species *Homo naledi* from the Dinaledi Chamber, South Africa. *eLife.* 4: 309561.

21. Clarey, T. L. 2017. Disposal of *Homo naledi* in a possible deathtrap or mass mortality scenario. *Journal of Creation.* 31 (2): 61-70.

22. Egeland, C. P. et al. 2018. Hominin skeletal part abundances and claims of deliberate disposal of corpses in the Middle Pleistocene. *Proceedings of the National Academy of Sciences.* 115 (18): 4601-4606.

23. Val, A. 2016. Deliberate body disposal by hominins in the Dinaledi Chamber, Cradle of Humankind, South Africa? *Journal of Human Evolution.* 96: 145-148.

24. Rupe, C. and J. Sanford. 2017. *Contested Bones.* Canandaigua, NY: FMS Publications, 155-178.

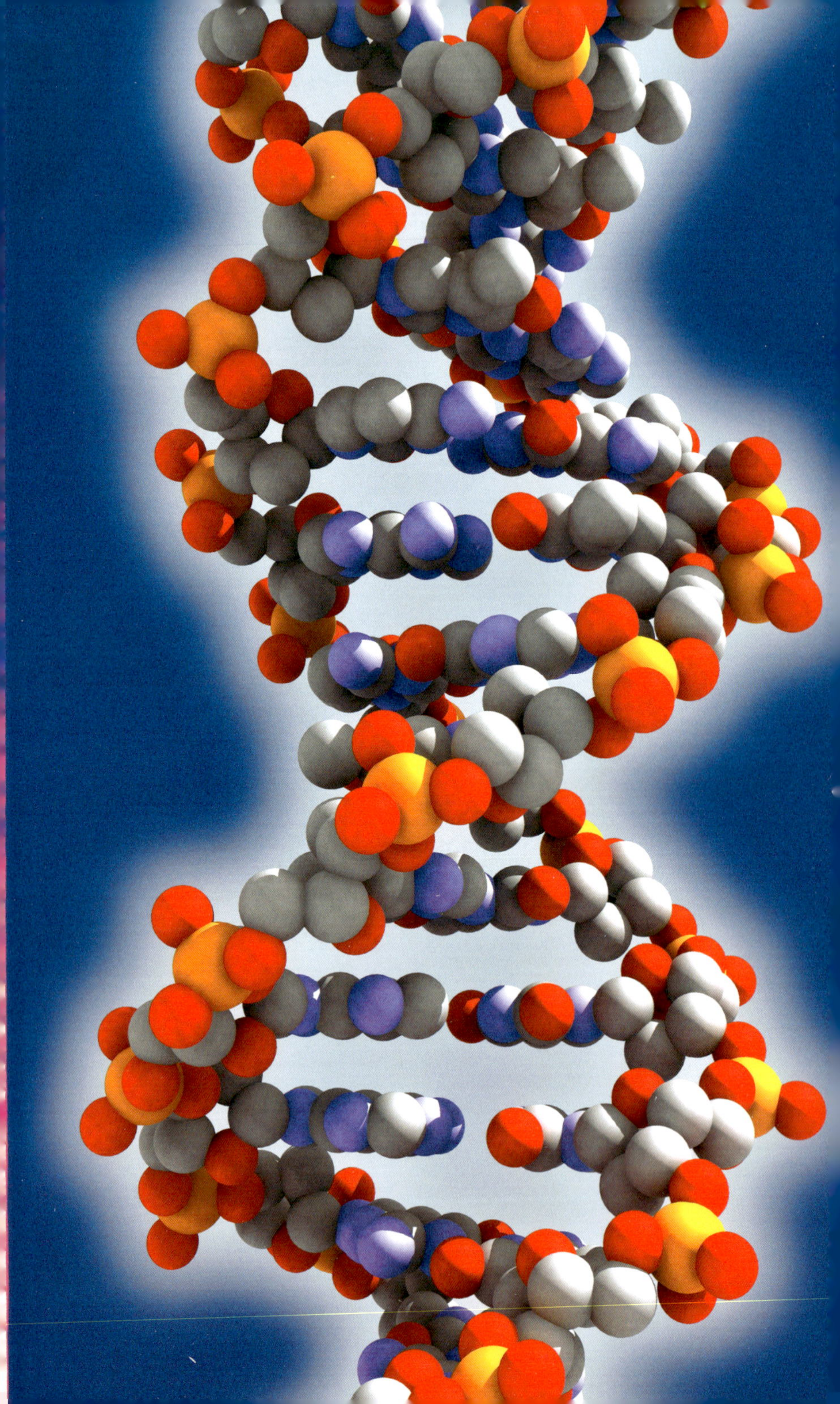

# 17

# HUMAN CHROMOSOME 2 FUSION NEVER HAPPENED

Jeffrey P. Tomkins, Ph.D.

- Evolutionists struggle to explain why humans have 46 chromosomes and apes have 48 if both descended from a common ancestor. The supposed answer is the fusion of two chromosomes in the past.

- The alleged fusion site isn't connected to satellite DNA sequence like documented fusions are, and it's too small and muddled to be the fusion of two chromosomes.

- Most importantly, the fusion site is located inside a gene and contains intricate coding functionality that refutes the theory of a fusion.

- The alleged cryptic centromere site is inside a large protein-coding gene, further refuting the fusion idea.

- The human chromosome 2 fusion theory doesn't hold up to scientific scrutiny.

A popular argument for apes evolving into humans is known as the chromosome fusion. The motivation for this concept is the evolutionary problem that apes have an extra pair of chromosomes—

humans have 46, while apes have 48. If humans evolved from an ape-like creature only three to six million years ago, then why do humans and apes have this discrepancy?

The evolutionary solution proposes that an end-to-end fusion of two small ape-like chromosomes (named 2A and 2B) produced human chromosome 2 (Figure 1). The concept of a fusion first arose in 1982 when scientists examined the similarities of human and ape chromosomes under a microscope. While the technique was somewhat crude, it was enough to start the idea.[1]

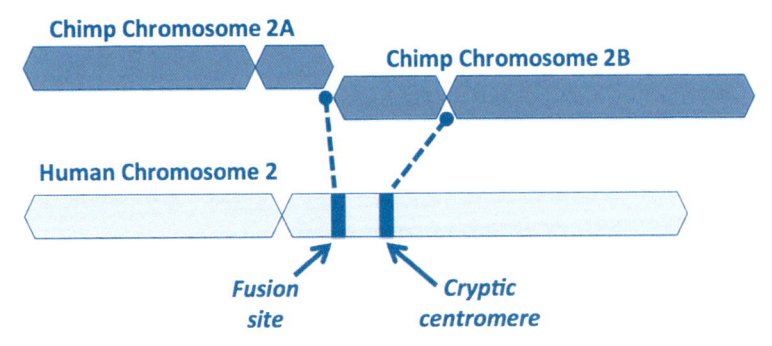

*Figure 1.* Hypothetical model in which chimpanzee chromosomes 2A and 2B fused end-to-end to form human chromosome 2. The chromosomes are drawn to scale according to cytogenetic images published by Yunis and Prakash.[1] Note the size discrepancy, which is about 10% or 24 million bases based on the known size of human chromosome 2.

## The So-Called Fusion Site

The first actual DNA signature of a possible fusion event was discovered in 1991 on human chromosome 2.[2] Researchers found a small, muddled cluster of telomere-like end sequences that vaguely resembled a possible fusion. A telomere is a six-base sequence of the DNA letters TTAGGG repeated over and over again at the ends of chromosomes.

However, the fusion signature was somewhat of an enigma based on the real fusions that occasionally occur in nature. All documented fusions in living animals involve a specific type of sequence called satellite DNA (satDNA) located in chromosomes and found in breakages and fusions.[3-5] The fusion signature on human chromosome 2 was missing this satDNA.[6]

Another problem is the small size of the fusion site, which is only 798 DNA letters long. Telomere sequences at the ends of chromosomes are 5,000 to 15,000 bases long. If two chromosomes had fused, you should see a fused telomere signature of 10,000 to 30,000 bases long—not 798.

Not only is the small size a problem for the fusion story, the signature doesn't represent a clear-cut fusion of telomeres. Figure 2 shows the DNA letters of the 798-base fusion site with the six-base (DNA letter) intact telomere sequences emphasized in bold

```
TGAGGGTGAGGGTTAGGGTTTGGGTTGGGTTTGGGGTTGGGGTTGGGGTAGGGGTGGGGTTGGGGTT
GGGGTTGGGGTTAGGGGTAGGGGTAGGGGTAGGGGTAGGGTCAGGGTCAGGGTCAGGGTTAGGGTT
TTAGGGTTAGGATTTTAGGGTTAGGGTAAGGGTTAAGGGTTGGGGTTGGGGTTAGGGTTAGGGGTT
AGGGTTGGGGTTGGGGTTGGGGTTGGGGTTGGGGTTAGGGTTAGCTAAACCTAACCCCTAAC
CCCTAACCCCAACCCCAACCCCAACCCTACCCCTACCCCTACCCCTAACCCCAACCCCCACCCTTAAC
CCTTAACCCTTACCCTAACCCTAACCCAAACCCTAACCCTACCCTAACCCTAACCCAACCCTAACCC
TAACCCTACCCTAACCCTAACACCCTAAAACCGTGACCCTGACCTTGACCCTGACCCTTAACCCTTAA
CCCTAACCATAACCCTAAACCCTAACCCTAAACCCTAACCCTAAACCCTAACCCTAACACTACCCT
ACCCTAACCCCAACCCCTAACCCCTAACCCTAACCCTACCCCTAACCCCAACCCCAGCCCCAACCCT
TACCCTAACCCTACCCTAACCCTTAACCCTAACCCCTAACCCTAACCCCTAACCCTAACCCTACCCC
AACCCCAAACCCAACCCTAACCCAACCCTAACCCCTAACCCTAACCCCTACCCTAACCCCTAGCCCT
AGCCCTAGCCCTAACCCTAACCCTCGCCCTAACCCTCACCCTAACCCTCACCCTCACCCTAA
```

***Figure 2.*** *The 798 sequence of the alleged fusion site. Intact forward (TTAGGG) and reverse complement (CCCTAA) telomere sequences are in bold font. The actual alleged point of fusion (AA) is underlined.*

print. When the fusion sequence is compared to a pristine fusion signature of the same size, it is only 70% identical overall.

Conventional researchers pointed out this discrepancy and labeled the fusion site as "degenerate."[7] Given the standard theoretical model of human evolution, it should be about 98 to 99% identical, not 70%. The researchers describing this discovery commented, "Head-to-head arrays of repeats at the fusion site have degenerated significantly (14%) from the near perfect arrays of (TTAGGG)$_n$ found at telomeres." They asked the pertinent question: "If the fusion occurred within the telomeric repeat arrays less than ~6 Mya, why are the arrays at the fusion site so degenerate?"[7] It should be noted that the 14% degeneration cited by the authors refers to the corruption of just the six-base sequences themselves, not the whole 798 bases.

**The Fusion Site Inside a Gene?**

The most remarkable anti-evolutionary finding about the fusion site turned out to be its location and what it actually does. This discovery happened while I was reading a detailed analysis of 614,000 bases of DNA sequence surrounding the alleged fusion site. I noticed that the fusion site was located *inside* a gene. Remarkably, this oddity wasn't acknowledged in the text of the paper.[8]

A finding like this is highly noteworthy. This piece of information would've been the nail in the coffin, so to speak, which might be why the researchers declined to discuss it. This major anomaly inspired me to give the fusion site a closer examination. The paper had been published in 2002, and I took notice of it in 2013. During that time, a huge amount of data on the structure and function of the human genome had been published, and likely much more to the story remained to be uncovered.

When I performed further research, I verified that the fusion site was positioned inside an RNA helicase gene now called *DDX11L2*. Most genes in plants and animals have their coding segments in pieces called exons so they can be alternatively spliced. Based on the addition or exclusion of exons, genes can produce a variety of products. The intervening regions between exons are called introns. They often contain a variety of signals and switches that control gene function. The alleged fusion site is positioned inside the first intron of the *DDX11L2* gene (Figure 3).[9]

The DNA molecule is double-stranded, with a plus strand and a minus strand. It was engineered this way to maximize information density while in-

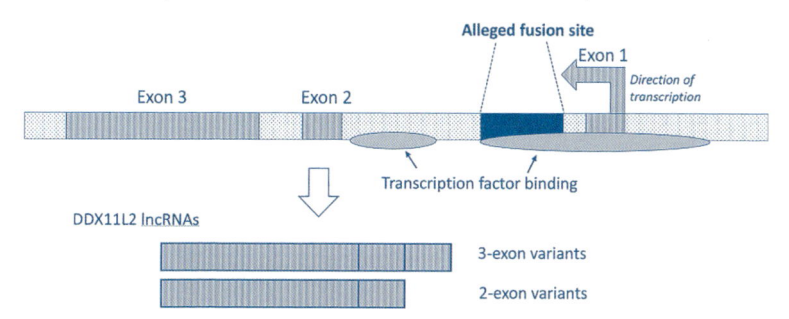

*Figure 3.* Simplified illustration of the alleged fusion site inside the first intron of the DDX11L2 gene. The graphic also shows two versions of short and long transcript variants produced, along with areas of transcription factor binding. The arrow in the first exon depicts the direction of transcription.

creasing efficiency and function. Genes run in different directions on the opposing strands. As it turns out, the *DDX11L2* gene is encoded on the minus strand. Genes in humans are like Swiss army knives and can produce a variety of RNAs. In the case of the *DDX11L2* gene, it produces short variants consisting of two exons and long variants with three (Figure 3).[9]

**The Fusion Site Is a Gene Promoter**

What might this *DDX11L2* gene be doing? My research showed that it's expressed in at least 255 different cell or tissue types.[9] It's also co-expressed (turned on at the same time) with a variety of other genes and is associated with cell signaling in the extracellular matrix and blood cell production. That the "fusion" sequence is located inside a highly functional gene refutes the idea that it's the accidental byproduct of a telomeric fusion. Functional genes are not formed by catastrophic chromosomal fusions!

Even more amazing is that the fusion site is functional and serves an important engineered purpose. The site acts as a switch for controlling gene activity. A wealth of biochemical data showed that 12 different proteins called transcription factors regulate this segment of the gene. One of these is RNA polymerase II, the main enzyme that copies RNA molecules from DNA in a process called transcription. Further supporting this discovery is that transcription begins inside the "fusion" site.

Technically, we would call the activity in the alleged fusion site a promoter region. Promoters are the main switches at the beginning of genes that turn them on. They are also where the RNA polymerase starts to create an RNA. Many genes have alternative promoters, like the *DDX11L2* gene.

The *DDX11L2* gene contains two areas of transcription factor binding. The first is in the promoter in front of the first exon. The second is in the first intron corresponding to the fusion site sequence. The *DDX11L2* gene is controlled, with the "fusion" sequence playing a key role, and the RNA transcripts produced are very intricate. The RNAs contain a wide variety of binding and control sites for a class of small regulatory molecules called microRNAs.[9]

**Functional Internal Telomere Sequences Are All Over the Genome**

Internally located telomere sequences are found all over the human genome. These seemingly out-of-place telomere repeats have been dubbed interstitial telomeres. The presence of these sequences presents another challenge for the fusion site idea—very few of the telomere repeats occur in the fusion site in tandem. As noted in Figure 2, the sequence of the 798-base fusion site contains only a few instances where two repeats are in tandem, and none have three repeats or more. However, there are many other interstitial telomere sites all over the human genome where the repeats occur in perfect tandem three to ten times or more.[10-11]

Besides their role at the ends of chromosomes, it appears that interstitial telomeric repeats may serve an important function in the genome related to gene expression. In one research project, I identified telomere repeats all over the human genome and intersected their genomic locations with diverse data sets containing functional biochemical information for gene activity.[12] I found thousands of internal telomeric repeats across the genome that were directly associated with the hallmarks of gene expression. The same type of transcription factor binding and gene activity occurring at the alleged fusion site was also occurring genome-wide at many other interstitial telomeric repeats. Clearly, these DNA features are not accidents of evolution but intelligently designed functional code.

**No Cryptic Centromere Inside a Gene**

Another problem with the fusion model is the lack of evidence for a signature of an extra centromere region. Centromeres are sections of chromosomes, often in central locations, that play key roles during cell division. As depicted in Figure 1, the newly formed chimeric chromosome would've had two centromere sites immediately following the alleged head-to-head fusion of the two chromosomes. In such a case, one of the centromeres would be functional while the other would be disabled. Two active centromeres are bad news for chromosomes and would lead to dysfunction and cell destruction.

The evidence for a cryptic (disabled) centromere on human chromosome 2 is even weaker than that for a telomere-rich fusion site. Evolutionists explain the lack of a distinguishable nonfunctional secondary centromere by arguing that a second centromere would've been rapidly eliminated by "natural selection." Since the disabled centromere was no longer useful in the genome, then no functional restraints were placed on it, causing it to deteriorate over time.

However, evidence for a second remnant centromere at any stage of sequence degeneration is problematic for evolution. Functional centromere sequences are composed of a repetitive type of DNA called alphoid sequences. Each alphoid repeat extends about 171 bases long. Some types of alphoid repeats are found all over the genome, while others are specific to centromeres. The structure of the sequences found at the cryptic centromere site on human chromosome 2 doesn't match those associated with functional human centromeres.[13] Even worse for the evolutionary model is that no highly similar counterparts can be found in the chimp genome—these sequences are human-specific.[13]

The alleged fossil centromere is exceptionally tiny compared to a real one. The size of a normal human centromere ranges in length between 250,000 and 5,000,000 bases.[14] The alleged cryptic centromere is only 41,608 bases long. It's also important to note that three different regions of it aren't even alphoid repeats.[15] Two of these are called retroelements, with one being a LPA3/LINE repeat 5,957 bases long and the other an SVA-E element with 2,571 bases. When we subtract the insertions of these non-alphoid sequences, it gives a length of only 33,080 bases, which is a fraction of the length of a real centromere.

But the most serious evolutionary problem with the idea of a fossil centromere is that it's positioned inside a gene just like the alleged fusion site. The alleged cryptic centromere is located inside the *ANKRD30BL* gene, and its sequence spans both intron and exon regions of the gene.[12,15]

In fact, the part of the "fossil" centromere sequence that lands inside an exon actually codes for amino acids in the resulting gene's protein. The gene produces a protein that's believed to be involved in the structural network of proteins called the cytoskeleton.[16] That the "fossil" or "cryptic" centromere is a functional region inside an important protein-coding gene completely refutes the idea that it's defective.

**Conclusion: No Fusion**

Due to the muddled signatures and small sizes of the alleged fusion and fossil centromere sites, it's highly questionable that their sequence was derived from an ancient chromosome fusion. Not only that, they represent functional sequence inside genes. The alleged fusion site is an important genetic switch called a promoter inside the *DDX11L2* long noncoding RNA gene. The "fossil" centromere contains both coding and noncoding sequences inside a large ankyrin repeat protein-coding gene.

This is an undeniable double refutation of the fusion idea. The overwhelming scientific conclusion is that the fusion never happened.

*References*

1. Yunis, J. J. and O. Prakash. 1982. The origin of man: a chromosomal pictorial legacy. *Science*. 215 (4539): 1525-1530.

2. Ijdo, J. W. et al. 1991. Origin of human chromosome 2: An ancestral telomere–telomere fusion. *Proceedings of the National Academy of Sciences*. 88 (20): 9051-9055.

3. Chaves, R. et al. 2003. Molecular cytogenetic analysis and centromeric satellite organization of a novel 8;11 translocation in sheep: a possible

intermediate in biarmed chromosome evolution. *Mammalian Genome.* 14 (10): 706-710.

4. Tsipouri, V. et al. 2008. Comparative sequence analyses reveal sites of ancestral chromosomal fusions in the Indian muntjac genome. *Genome Biology.* 9 (10): R155.

5. Adega, F., H. Guedes-Pinto, and R. Chaves. 2009. Satellite DNA in the Karyotype Evolution of Domestic Animals—Clinical Considerations. *Cytogenetic and Genome Research.* 126 (1-2): 12-20.

6. Tomkins, J. P. and J. Bergman. 2011. Telomeres: implications for aging and evidence for intelligent design. *Journal of Creation.* 25 (1): 86-97.

7. Fan, Y. et al. 2002. Genomic Structure and Evolution of the Ancestral Chromosome Fusion Site in 2q13–2q14.1 and Paralogous Regions on Other Human Chromosomes. *Genome Research.* 12 (11): 1651-1662.

8. Fan, Y. et al. 2002. Gene Content and Function of the Ancestral Chromosome Fusion Site in Human Chromosome 2q13–2q14.1 and Paralogous Regions. *Genome Research.* 12 (11): 1663-1672.

9. Tomkins, J. P. 2013. Alleged Human Chromosome 2 "Fusion Site" Encodes an Active DNA Binding Domain Inside a Complex and Highly Expressed Gene—Negating Fusion. *Answers Research Journal.* 6: 367-375.

10. Azzalin, C. M., S. G. Nergadze, and E. Giulotto. 2001. Human intrachromosomal telomeric-like repeats: sequence organization and mechanisms of origin. *Chromosoma.* 110: 75-82.

11. Ruiz-Herrera, A. et al. 2008. Telomeric repeats far from the ends: mechanisms of origin and role in evolution. *Cytogenetic and Genome Research.* 122 (3-4): 219-228.

12. Tomkins, J. P. 2018. Combinatorial genomic data refute the human chromosome 2 evolutionary fusion and build a model of functional design for interstitial telomeric repeats. In *Proceedings of the Eighth International Conference on Creationism.* J. H. Whitmore, ed. Pittsburgh, PA: Creation Science Fellowship, 222-228.

13. Tomkins, J. and J. Bergman. 2011. The chromosome 2 fusion model of human evolution—part 2: re-analysis of the genomic data. *Journal of Creation.* 25 (2): 111-117.

14. Aldrup-Macdonald, M. E. and B. A. Sullivan. 2014. The Past, Present, and Future of Human Centromere Genomics. *Genes (Basel).* 5 (1): 33-50.

15. Tomkins, J. P. 2017. Debunking the Debunkers: A Response to Criticism and Obfuscation Regarding Refutation of the Human Chromosome 2 Fusion. *Answers Research Journal.* 10: 45-54.

16. Voronin, D. A. and E. V. Kiseleva. 2008. Functional Role of Proteins Containing Ankyrin Repeats. *Cell and Tissue Biology.* 49 (12): 989-999.

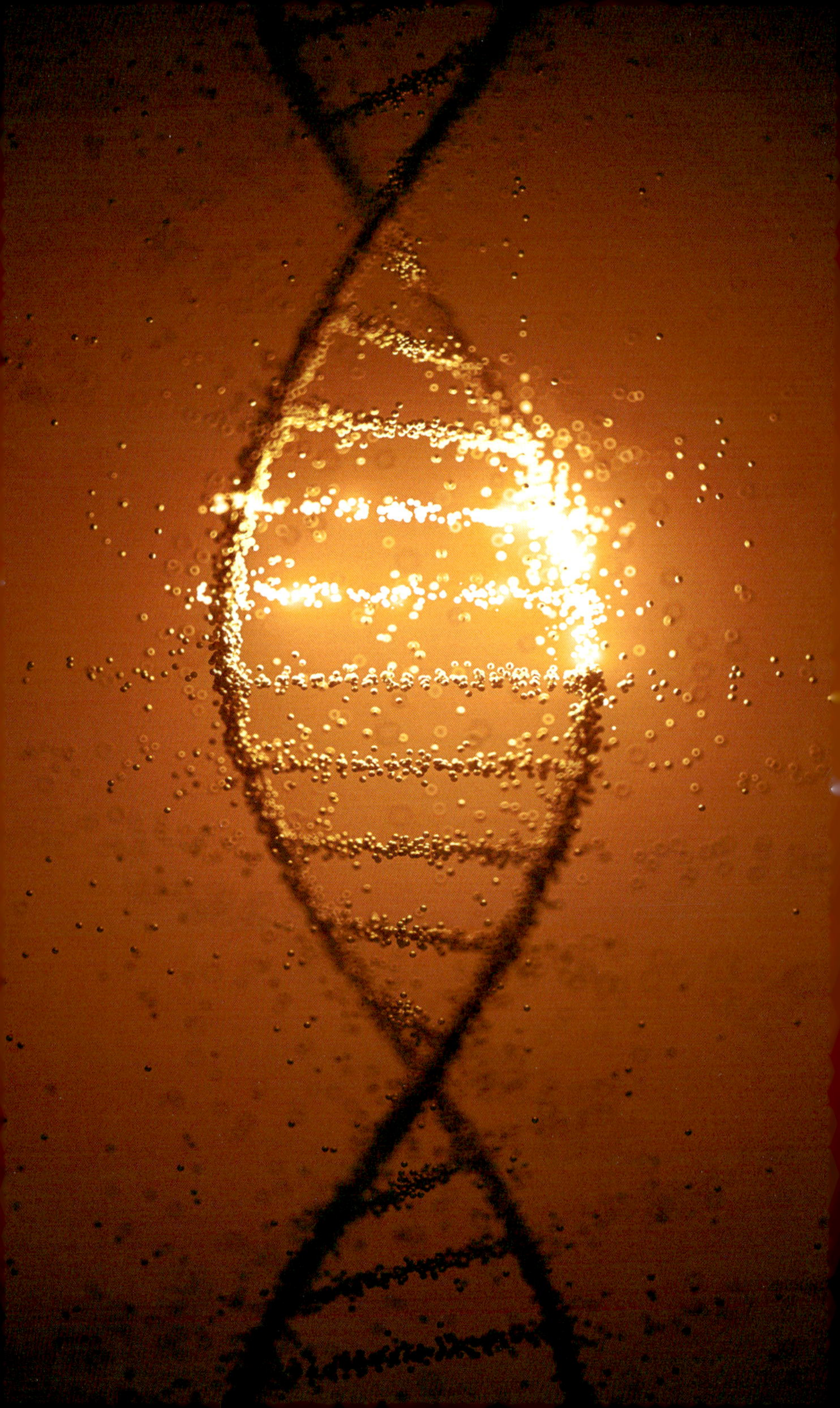

# 18

# HUMAN GENOME 20TH ANNIVERSARY— JUNK DNA HITS THE TRASH

Jeffrey P. Tomkins, Ph.D.

The first rough drafts of the human genome were reported in 2001 (one in the private sector and one in the public sector).[1,2] Since then, after decades of intensive globally conducted research, the data have revealed a wealth of complexity that completely upset many evolutionary misconceptions.[3] Most importantly, the idea of "junk DNA" was debunked in favor of a new model, one containing pervasive functionality and network complexity. This complexity is only beginning to be revealed—an inconvenient fact that points to an omnipotent Creator.

A 2021 cover story in the journal *Nature* summarized the first 20 years after the first drafts of the human genome were published.[3] When the first phase of research was completed in 2001, scientists discovered that the genome contained about 25,000 protein-coding genes and that the actual coding segments of these genes only accounted for about 2% of the total DNA sequence.

Many evolutionists found affirmation in these initial reports. This was because evolutionary theory predicted that there should be vast regions of "junk

DNA" in the human genome. These alleged nonfunctional regions would then be randomly churning out new genes for nature to magically select.[4,5] But this misguided evolutionary speculation was short-lived.

After 2001, numerous research projects demonstrated that these uncharted and mysterious regions of the human genome were not junk at all. Rather, they were vital to life and good health. In a subsection of a *Nature* article entitled "Not Junk," the authors say:

> With the HGP [human genome project] draft in hand, the discovery of non-protein-coding elements exploded. So far, that growth has outstripped the discovery of protein-coding genes by a factor of five, and shows no signs of slowing.[3]

They also said:

> Thanks in large part to the HGP, it is now appreciated that the majority of functional sequences in the human genome do not encode proteins. Rather, elements such as long non-coding RNAs, promoters, enhancers and countless gene-regulatory motifs work together to bring the genome to life.[3]

The main points of the first 20 years of research on the human genome can be summarized as follows.

1) The human genome is a complete storehouse of important information, and this negates the concept of junk DNA.

2) Protein-coding genes are largely a basic set of instructions within a complex and larger repertoire of regulatory DNA sequence.

3) Many more genes exist (compared to protein-coding genes) that code for functional RNA molecules that are not used to make proteins but do other jobs in the cell.

4) Many regulatory switches and control features exist in the human genome that regulate its function.

The pervasive and complex design of the human genome is exactly what the Bible teaches. After all, the Scriptures say in Psalm 139:14, "I will praise You, for I am fearfully and wonderfully made; Marvelous are Your works, And that my soul knows very well."

*References*

1. Venter, J. C. et al. 2001. The Sequence of the Human Genome. *Science*. 291 (2001): 1304-1351.
2. International Human Genome Sequencing Consortium. 2001. Initial Sequencing and Analysis of the Human Genome. *Nature*. 409 (2001): 860-921.
3. Gates, A. J., D. M. Gysi, M. Kellis, and A. L. Barabási. 2021. A wealth of discovery built on the Human Genome Project—by the numbers. *Nature*. 590: 212-215.
4. Tomkins, J. P. 2017. Evolutionary Clock Futility. *Acts & Facts*. 46 (3): 16.
5. Tomkins, J. P. and J. Bergman. 2015. Evolutionary molecular genetic clocks—a perpetual exercise in futility and failure. *Journal of Creation*. 29 (2): 26-35.

*Machu Picchu, Peru*

# 19

# HUMAN HIGH-ALTITUDE HABITATION REVEALS ADAPTIVE DESIGN

Jeffrey P. Tomkins, Ph.D.

Humans have the remarkable ability to inhabit high altitudes where living conditions are harsh and challenging. A study in *Genome Biology and Evolution* showed that specific epigenetic modifications are important in forming this ability.[1] These results refute chance evolution—revealing innate systems of complex biological engineering that undergird high-altitude adaptation.

Humans have colonized an amazing array of challenging environments across the earth, from arid deserts to frozen tundra, tropical rainforests, and some of the highest mountain regions. Among them, high-altitude mountain living is one of the most challenging. Nevertheless, approximately 2% of the world's people permanently inhabit high-altitude regions of over 2,500 meters (1.5 miles) above sea level.

In these places, oxygen is sparse, ultraviolet radiation is high, and temperatures are low. Examples of people groups that live at these extreme altitudes include native Andeans, Tibetans, Mongolians, and Ethiopians. Studies of Andeans and Tibetans have re-

*Pitumarca Cuzco, Peru, Andes Mountains of South America*

vealed an increase in chest circumference (associated with greater lung volume), elevated oxygen saturation, and a low hypoxic ventilatory response. This is contrasted with lowlanders who travel to high altitudes and experience an intense hypoxic ventilatory in response to the lower levels of oxygen.[1,2]

For years, biologists tried to find a genetic component to this unique suite of adaptive high-altitude responses—but with only moderate success. Previous studies showed a propensity for a few different gene variants to exist at higher frequencies in high-altitude populations. But nothing definitive has been found regarding a significant and consistent heritable genetic mechanism that would explain the unique changes in physiology and anatomy. As one conventional author stated, "The underlying mechanism for this remains poorly understood."[3]

However, the *Genome Biology and Evolution*

study revealed for the first time an exciting feature of built-in adaptive design based on the broadening field of epigenetics. While epigenetics is a diverse field of research, one popular sector within it studies the addition of methyl molecule tags that are strategically placed on cytosine bases throughout the DNA in and around genes to regulate their expression. This process is known as methylation. These methyl tags do not change the DNA sequence itself, but rather modify how the DNA is used, resulting in diverse bodily changes. Furthermore, these methylation patterns can be inherited for several generations so that succeeding populations are primed for the specific environment they will live in.

In this research, the authors compared the DNA methylation patterns of a specific people group of highland Quechua ancestry in the Andes region of South America. The study included people who lived at high altitudes, low altitudes, and those who migrated from low to high. First, the researchers discovered that many of these adaptive epigenetic modifications were associated with genes involved in red blood cell production, glucose metabolism, and skeletal muscle development. These are standard gene regions connected to high-altitude adaptation.

Second, the researchers found that the whole process of epigenetic-based adaptation to high altitudes was triggered early during embryonic development. Children were born fully adapted with bigger lungs and other important needed traits. This phenomenon is known as developmental adaptation or adaptive plasticity. In other words, many developmental sensors detect key features of the environment. This information is processed by cellular machinery that then modifies the organism's genome with strategically placed methyl tags. The methyl tags properly control the development of a high-altitude-adapted baby.

While evolutionists have traditionally regarded high-altitude adaptation in humans as evidence of chance evolution, this study exposes the futility of the Darwinian paradigm. Darwinism purports that nature has the volition and capability to select beneficial traits based on a set of DNA mutation options. In reality, the environment (nature) merely represents a set of parameters (temperature, oxygen content, etc.) that engineered living systems detect and track through elaborate sensors. These data are then processed through complex internal systems that provide specific outputs and solutions—resulting in adaptation.

This study reiterates that we should do biology as if Charles Darwin had never existed. Creatures adapt to diverse environments because they were fabulously designed and engineered by an omnipotent Creator—the Lord Jesus Christ.

*References*

1. Childebayeva A. et al. 2021. Genome-Wide Epigenetic Signatures of Adaptive Developmental Plasticity in the Andes. *Genome Biology and Evolution.* 13 (2): evaa239.

2. Thomas, B. Highlander Tibetans Show Adaptation, Not 'Natural Selection.' *Creation Science Update.* Posted on ICR.org July 15, 2010, accessed March 2, 2021.

3. McGrath, C. 2021. Highlight: The Epigenetics of Life at 12,000 ft. *Genome Biology and Evolution.* 13 (2): evaa266.

# 20

## HUMANS AND NEANDERTHALS ARE MORE SIMILAR THAN POLAR AND BROWN BEARS

Jeffrey P. Tomkins, Ph.D.

A study led by Oxford University researchers confirms that Neanderthals and humans were genetically similar and often interbred. They were even closer genetically than polar and brown bears are to each other—bears that are known to mate and produce viable offspring in the wild.[1] Along with a plethora of previous DNA studies, this research further confirms that Neanderthals were an ancient group of humans descended from Noah's three sons and their wives after the global Flood.[2-4]

Analyzing ancient DNA demonstrated that anatomically modern humans mated with their close archaic relatives, Neanderthals and Denisovans. However, many evolutionists still claim that modern humans and their archaic cousins were radically different and that their ability to breed and produce viable offspring was at the edge of biological compatibility.

In this study, the researchers developed a genetic distance metric to predict the fertility of the first generation of hybrid offspring between the mating of any two mammalian species. They did this by analyzing genetic sequences from different mammal species that were already known to successfully mate. By correlating genetic distance with offspring fertility, they showed that the greater the genetic distance, the less likely it would be that the offspring would be fertile. Then the researchers effectively used the genetic distance values to determine thresholds of fertility for various mammals.

When the genetic distance between humans, Neanderthals, and Denisovans was calculated, the values were even closer than those found between various animals that are known to readily and easily hybridize in the wild. This includes polar bears mating with brown bears and coyotes mating with wolves. Professor Greger Larson at Oxford University, the senior author of the study, said, "Humans and Neanderthals and Denisovans were able to produce live fertile young with ease."[5]

These findings confirm that Neanderthals were fully human. Creationists believe that they likely lived shortly after the global Flood since their remains are typically found in ritually buried graves in cave systems.[6] And not only is the DNA of Neanderthals identical to modern humans, but the so-called "archaic traits" of a pronounced brow ridge and a sloping forehead can still be found in modern humans living today.[7] Once again, the facts of science confirm the Scriptures—not the mythical story of human evolution.

*References*

1. Allen, R. et al. 2020. A mitochondrial genetic divergence proxy predicts the reproductive compatibility of mammalian hybrids. *Proceedings of the Royal Society*. 287 (1928).
2. Tomkins, J. P. DNA Proof That Neandertals Are Just Humans. *Creation Science Update*. Posted on ICR.org February 21, 2014, accessed June 15, 2020.
3. Tomkins, J. P. 2014. Ancient Human DNA: Neandertals and Denisovans. *Acts & Facts*. 43 (3): 9.
4. Tomkins, J. P. Neanderthal DNA Muddles Evolutionary Story. *Creation Science Update*. Posted on ICR.org April 7, 2020, accessed June 15, 2020.
5. Humans and Neanderthals: less different than polar and brown bears. University of Oxford News and Events. Posted on ox.ac.uk June 3, 2020, accessed June 15, 2020.
6. Clarey, T. L. 2020. Compelling Evidence for an Upper Cenozoic Flood Boundary. *Acts & Facts*. 49 (5): 9.
7. Tomkins, J. P. 2019. Recent Humans with Archaic Features Upend Evolution. *Acts & Facts*. 48 (4): 15.

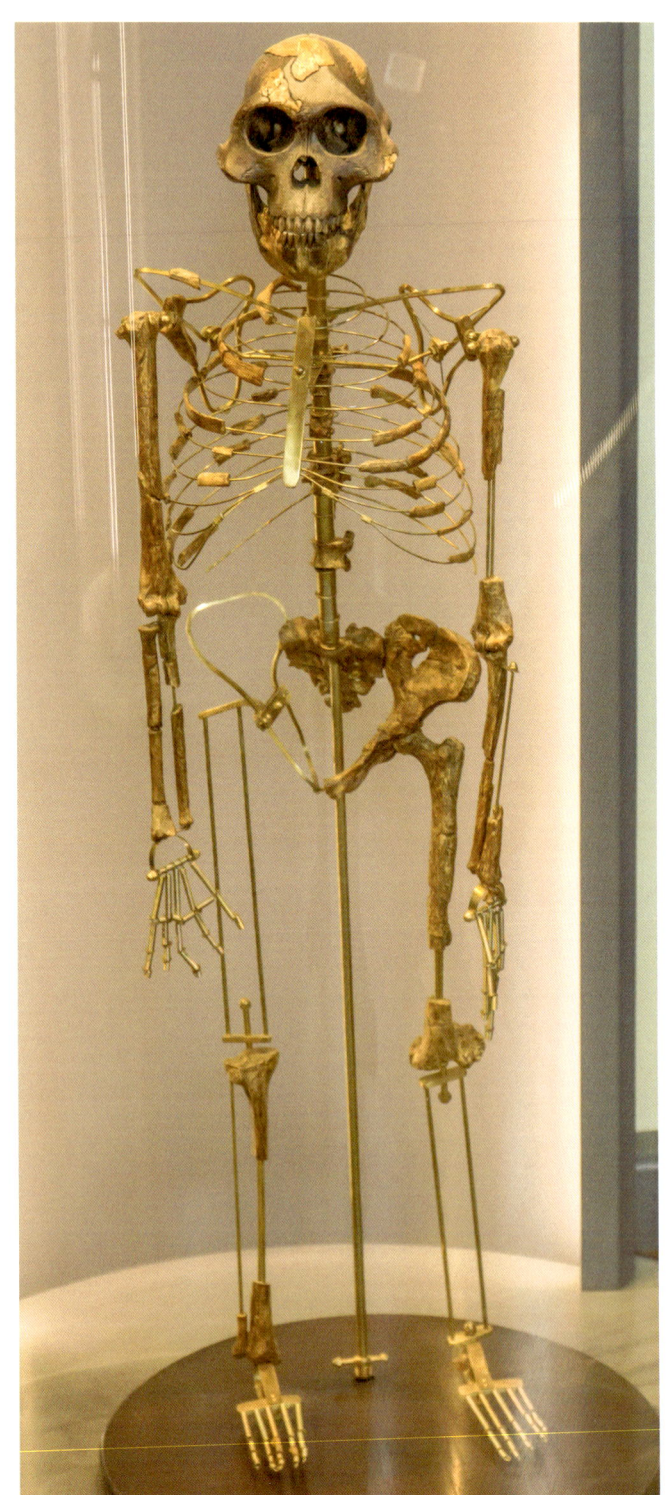

*Lucy*

# 21

## LUCY LANGUISHES AS A HUMAN-APE LINK

Frank Sherwin, D.Sc. (Hon.)

Human evolution has consistently lacked scientific and biblical merit. Although a parade of supposed transitions is displayed in every conceivable outlet, non-Darwinists maintain that the links between people and our alleged ape-like ancestors are *missing*.

Perhaps the most popular evolutionary relative of humans is *Australopithecus afarensis*—better known as Lucy.[1] Several hundred pieces of fossilized bone were discovered in east Africa in 1974. Evolutionists dated Lucy at 3.75 to 5 million years old and said her kind "probably evolved directly from [*Australopithecus*] *anamensis*."[2]

How should creationists respond to this compelling creature that supposedly links us to non-human ancestors? To begin with, many people are not aware of the subjectivity that is involved with piecing together shattered fossil remains. For example, two evolutionists stated in regard to Lucy:

> The sacrum and the auricular region of the ilium are shattered into numerous small fragments, such that the original form is difficult to elucidate. Hence it is not surprising that

the reconstructions by Lovejoy and Schmid show marked differences.[3]

These bone pieces have no dates on them, and no one can be sure they are from the same individual. The pieces fit where the biased researcher would like them to go. Whatever fragments are missing must be filled in with plaster of Paris and imagination.[4] This is certainly true with Lucy. Her bones are what would be expected on the basis of creation: "Lucy's fossil remains match up remarkably well with the bones of a pygmy chimp."[5]

Anatomical evidence shows this creature was ape-like with a nonhuman gait.[6] She knuckle-walked and climbed trees. But visitors to natural history museums see a reconstructed creature with an erect human-like posture. These displays show Lucy with an intelligent stare and skin color and hair—added by an evolutionary artist who never saw her when she was alive and must guess the majority of her reconstructed features. Humans have a U-shaped mandible, but Johanson agreed that Lucy's mandible (jaw) was V-shaped or ape-like.

> Her jaw was the wrong shape....I interpreted other things in her dentition [teeth] as primitive also, as pointing away from the human condition and back in the direction of apes.... The larger jaws had some of those same primitive features.[7]

Despite the evidence, Johanson was determined to see Lucy as a human ancestor. His bias is revealed

*Artist's interpretation of* Australopithecus afarensis

by his interpretation of a single arm bone he discovered in the sand. He stated:

> This time I knew at once I was looking at a hominid elbow. I had to convince Tom [Gray], whose first reaction was that it was a monkey's. But that wasn't hard to do.[8]

Other fossil fragments fail to fit the evolutionary picture. Lucy's shoulder blade was "virtually identical to that of a great ape and had a probability less than 0.001 of coming from the population represented by our modern human sample."[9]

Despite displays at multiple museums and zoos, Lucy had short, curved toe and finger bones.[10]

Humans have no curvature in these bones. In addition, Lucy had a locking hand joint, while people are designed with a non-locking hand joint.

> A chance discovery made by looking at a cast of the bones of "Lucy," the most famous fossil of *Australopithecus afarensis*, shows her wrist is stiff, like a chimpanzee's, Brian Richmond and David Strait of George Washington University in Washington, D.C., reported. This suggests that her ancestors walked on their knuckles.[11]

Later, it was reported that Lucy "had an exceptionally powerful upper body, thanks to spending a lot of time climbing trees."[12]

Her brain case was the same size as the common chimpanzee. Humans have a barrel-shaped rib cage, but Lucy's rib cage was conical-shaped (as found in apes). Paleontologist Peter Schmid stated:

> When I started to put the skeleton together, I expected it to look human. Everyone had talked about Lucy as being very modern, very human, so I was surprised by what I saw. I noticed that the ribs were more round in cross-section, more like what you see in apes. Human ribs are flatter in cross-section. But the shape of the rib cage itself was the biggest surprise of all. The human rib cage is barrel shaped, and I just couldn't get Lucy's ribs to fit this kind of shape. But I could get them to make a conical-shaped rib cage, like what you see in apes.[13]

Returning to the pelvis, evolutionists determined that Lucy was more likely a male due to the pelvis being heart-shaped and without ridges.

> Consequently, there is more evidence to suggest that AL288-1 was male rather than female. A female of the same species as AL288-1 would have had a pelvis with a larger sagittal diameter and a less protruding sacral promontorium....It would perhaps be better to change the trivial name to "Lucifer" according to the old roman god who brings light after the dark night, because with such a pelvis "Lucy" would apparently have been the last of her species.[14]

Why are hominids like Lucy depicted as walking upright? Even evolutionists admit that "the origin of bipedalism, the key event in the evolution of hominids, remains a mystery."[15] Regardless, many evolutionists still believe Lucy walked upright, which they consider evidence of being an ancestor of modern humans. But is upright walking the final litmus test for chimp-to-man transition? The modern bonobo monkey walks upright perhaps 10%

*Lucy*

of the time, but this hardly means it is our ancestor. In fact, there has been some disagreement over how "modern" the bipedalism of Lucy was.[16] One study suggested:

> Even when Lucy walked upright, she may have done so less efficiently than modern humans, limiting her ability to walk long distances on the ground.[17]

A decade after Lucy's discovery, evolutionist Charles Oxnard showed the issue of whether she walked or not to be irrelevant to any human evolutionary story.

> The australopithecines…are now irrevocably removed from a place in the evolution of human bipedalism, possibly from a place in a group any closer to humans than to African apes and certainly from any place in the direct human lineage.[18]

*Australopithecus*, the group to which Lucy belongs, means "southern ape," and creationists believe that's exactly what these dozens of fossilized bone pieces were. As far as can be determined, Lucy was an extinct ape. People, on the other hand, were created in God's image, just as the Bible says (Genesis 1:27).

*References*

1. Johanson, D. C. and K. Wong. 2010. *Lucy's Legacy: The Quest for Human Origins.* New York: Crown Publishing Group.
2. Solomon, E. P., L. R. Berg, and D. W. Martin. 2011. *Biology*, 9th ed. Belmont, CA: Brooks/Cole, 473.
3. Häusler, M. and P. Schmid. 1995. Comparison of the Pelves of Sts 14 and AL288-1: Implications for Birth and Sexual Dimorphism in Australopithecines. *Journal of Human Evolution.* 29 (4): 363-383.

4. Evolutionary anthropologist Peter Dodson states in a video display in the Perot Museum in Dallas, "I have to tell you that imagination is a very, very important trait for paleontologists."
5. Zihlman, A. 1984. Pygmy chimps, people, and the pundits. *New Scientist*. 104 (1430): 39-40.
6. Collard, M. and L. C. Aiello. 2000. Human evolution: From forelimbs to two legs. *Nature*. 404 (6776): 339-340. See also Oliwenstein, L. 1995. Lucy's Walk. *Discover*. 16 (1): 42.
7. Johanson, D. C. and M. A. Edey. 1981. *Lucy: The Beginnings of Humankind*. New York: Simon & Schuster, 258.
8. Johanson, D., L. Johanson, and B. Edgar. 1994. *Ancestors: In Search of Human Origins*. New York: Villard Books, 60. See also Kimbel, W. H., D. C. Johanson, and Y. Rak. 1994. The first skull and other new discoveries of *Australopithecus afarensis* at Hadar, Ethiopia. *Nature*. 368 (6470): 449-451.
9. Susman, R. L., J. T. Stern, and W. L. Jungers. 1984. Arboreality and bipedality in the Hadar hominids. *Folia Primatologica*. 43 (2-3): 120-121.
10. Stern, J. T. and R. L. Susman. 1983. The locomotor anatomy of *Australopithecus afarensis*. *American Journal of Physical Anthropology*. 60 (3): 279.
11. Fox, M. Man's Early Ancestors Were Knuckle Walkers. *San Diego Union Tribune*, Quest Section, March 29, 2000.
12. Barras, C. Early hominin Lucy had powerful arms from years of tree-climbing. *New Scientist*. Posted on newscientist.com November 30, 2016.
13. Leakey, R. and R. Lewin. 1992. *Origins Reconsidered: In Search of What Makes Us Human*. New York: Anchor Books, 193-94.
14. Häusler and Schmid, Comparison of the Pelves of Sts 14 and AL288-1, 380.
15. Raven, P. H. et al. 2014. *Biology*, 10th ed. Dubuque, IA: McGraw-Hill, 724.
16. Kimbel, W. H. and L. K. Delezene. 2009. "Lucy" Redux: A Review of Research on *Australopithecus afarensis*. *Yearbook of Physical Anthropology*. 52: 2-48.
17. Human Ancestor 'Lucy' Was a Tree Climber. John Hopkins Medicine news release. Posted on hopkinsmedicine.org November 30, 2016, accessed March 13, 2017.
18. Oxnard, C. E. 1984. *The Order of Man: A Biomathematical Anatomy of the Primates*. New Haven: Yale University Press, 332.

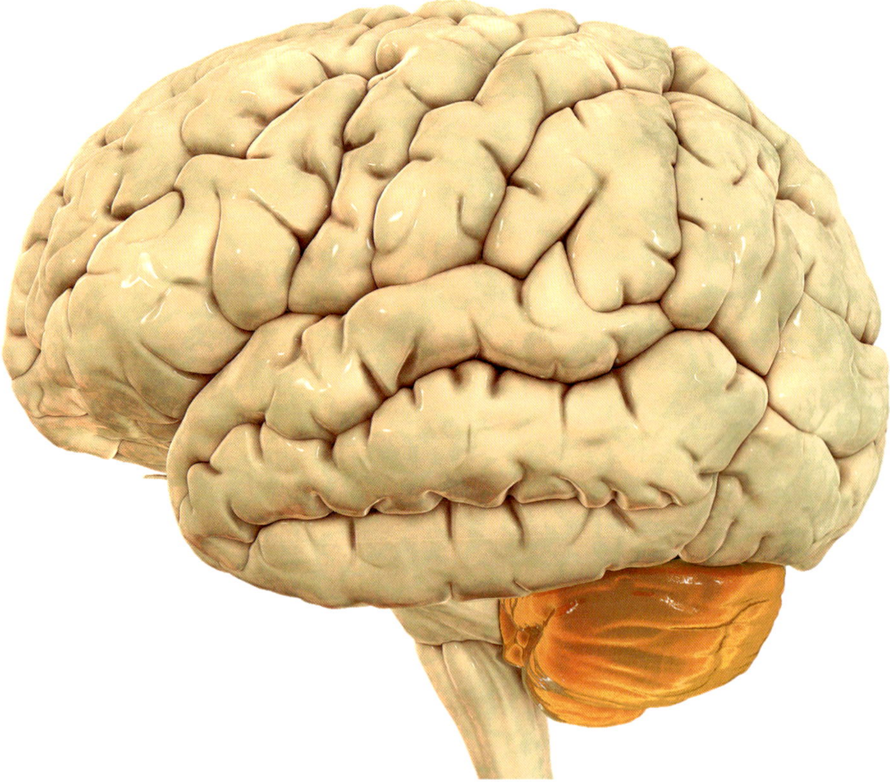

# 22

## MAN: SMART FROM THE START
Frank Sherwin, D.Sc. (Hon.)

- Scientists discovered that the cerebral cortex of the human brain has much more surface area than previously thought.
- While the discovery is factual, the evolutionary explanation has no evidence to support it.
- Creation scientists believe humans were distinctly created in the image of God from the beginning, so it makes perfect sense that humans have more brain space for complex thinking and function.

People were created with a three-pound brain that scientists will never fully understand. Evolutionists have tried to trace the evolution of the human neurological system (including the brain and spinal cord) from supposedly "lower life forms" but have been unsuccessful. There is a clear absence of evidence, by their own admissions, such as "surprisingly, little is known about the evolutionary origin of central nervous systems;"[1] "the origins of neural systems remain unresolved;"[2] and "when and how the animal nervous system arose has remained murky."[3]

One of the three major parts of our brain is the cerebellum, long known to be involved in coordination, regulation of balance, and other motor activities. The cerebellum contributes to our five senses, sits close to the brain stem, and is called by some the "little brain." Research has shown that the human cerebellum is more complex than scientists realized.[4]

> Until now, the cerebellum was thought to be involved mainly in basic functions like movement, but its expansion over time and its new inputs from cortical areas involved in cognition suggest that it can also process advanced concepts like mathematical equations.[5]

In their investigation, evolutionists compared human and macaque brains. The macaque is an Old World monkey.

> An SDSU neuroimaging expert discovered the tightly packed folds [of the cerebellum] actually contain a surface area equal to 80% of the cerebral cortex's surface area. In comparison, the macaque's cerebellum is about 30% the size of its cortex.[5]

The researchers revealed new information regarding the human cerebellum, but then they attempted to put an evolutionary spin on such discoveries.

> "The fact that it has such a large surface area speaks to the evolution of distinctively human behaviors and cognition," said Martin Sereno, psychology professor, cognitive neuroscientist and director of the SDSU MRI Imaging Center. "It has expanded so much that the folding patterns are very complex."[5]

On the other hand, creationists argue that the large surface area of the cerebellum speaks of God's design of distinctively human behaviors and cognition. Furthermore, since there is no evidence of people evolving from ape-like ancestors, we would say the human cerebellum was designed with very complex folding patterns from the beginning.

Who wouldn't applaud ongoing scientific research revealing the mysteries and complexities of the brain? But the evidence does not compel one to draw unscientific connections between people and monkeys. For evolutionists, the only option is to embrace a nonbiblical worldview despite the scientific evidence.

Sergio Almécija, a senior research scientist in the American Museum of Natural History's Division of Anthropology, said, "When you look at the narrative for hominin [bipedal apes including modern humans] origins, it's just a big mess—there's no consensus whatsoever."[6]

The evidence pushes us toward the truth—the Lord Jesus created people as people and apes as apes from the beginning.

*References*

1. Arendt, D. et al. 2008. The evolution of nervous system centralization. *Philosophical Transactions of the Royal Society B*. 363 (1496): 1523-1528.
2. Moroz, L. L. et al. 2014. The ctenophore genome and the evolutionary origins of neural systems. *Nature*. 510: 109-114.
3. Pennisi, E. 2019. The gluey tentacles of comb jellies may have revealed when nerve cells first evolved. *Science*. 363: 6424.
4. Sereno, M. I. et al. 2020. The human cerebellum has almost 80% of the surface area of the neocortex. *Proceedings of the National Academy of Sciences*. 117 (32): 19538-19543.

5. Nagappan, P. 'Little Brain' or Cerebellum Not So Little After All. San Diego State University news release. Posted on newscenter.sdsu.edu July 31, 2020.
6. Review: Studying Fossil Apes Key to Human Evolution Research. American Museum of Natural History press release. Posted on amnh.org May 6, 2021, accessed June 4, 2021.

## CONTRIBUTORS

Dr. Jeffrey Tomkins is Director of Research at the Institute for Creation Research and earned his Ph.D. in genetics from Clemson University.

Dr. Frank Sherwin is Science News Writer at the Institute for Creation Research. He earned an M.A. in zoology from the University of Northern Colorado and received an Honorary Doctorate of Science from Pensacola Christian College.

Dr. Brian Thomas is Research Scientist at the Institute for Creation Research and earned his Ph.D. in paleobiochemistry from the University of Liverpool.

Dr. Timothy Clarey is Research Scientist at the Institute for Creation Research and earned his Ph.D. in geology from Western Michigan University.

# IMAGE CREDITS

**t: top, b: bottom**

Animalparty / Wikimedia Commons: 84

Demin Alexei Barnaul / Wikimedia Commons: 46

Bigstock Photos: cover, 1, 3, 8, 13, 18t, 22, 26, 30, 38, 52-53, 66, 68, 72, 96, 108, 114, 128

Emőke Dénes / Wikimedia Commons: 10, 75

Ephraim33 / Wikimedia Commons: 18b, 125

Hawks et al / Wikimedia Commons: 62, 86

Johannes Maimillian / Wikimedia Commons: 120

Tila Monto / Wikimedia Commons: 34

Cicero Moraes / Wikimedia Commons: 123

Phiston / Wikimedia Commons: 19

Kameraad Pjotr / Wikimedia Commons: 60

Public Domain: 50, 112

Ryan Somma / Wikimedia Commons: 77

James St. John / Wikimedia Commons: 64

Jeffrey P. Tomkins: 98-99, 101

Dongju Zhang / Wikimedia Commons: 44

# THIS BOOK WAS ADAPTED FROM THE FOLLOWING MATERIALS.

Tomkins, J. P. 2019. A Literal Adam Is a Gospel Issue. *Acts & Facts.* 48 (6): 15.

Tomkins, J. P. 2018. Separate Studies Converge on Human-Chimp DNA Dissimilarity. *Acts & Facts.* 47 (11): 9.

Thomas, B. 2020. How Do Hominids Fit with the Bible? *Acts & Facts.* 49 (5): 20.

Tomkins, J. P. 95% of Human Genome Can't Evolve. *Creation Science Update.* Posted on ICR.org October 25, 2018.

Sherwin, F. 2021. Another Function of "Junk DNA" Discovered. *Acts & Facts.* 50 (10): 16.

Tomkins, J. P. Ape Spit Radically Different from Human. *Creation Science Update.* Posted on ICR.org November 12, 2019.

Clarey, T. and J. P. Tomkins. *Australopithecus* Ate Like an Ape. *Creation Science Update.* Posted on ICR.org August 20, 2019.

Tomkins, J. P. 2018. Codons Are Not Degenerate After All. *Acts & Facts.* 47 (7): 15.

Tomkins, J. P. Denisovan Epigenetics Reveals Human Anatomy. *Creation Science Update.* Posted on ICR.org October 10, 2019.

Tomkins, J. P. Denisovan DNA Shown to Be Human…Again. *Creation Science Update.* Posted on ICR.org April 9, 2018.

Tomkins, J. P. 2017. DNA Science Disproves Human Evolution. *Acts & Facts.* 46 (6): 10-13.

Sherwin, F. 2021. Does Recent Research Support Human Evolution? *Acts & Facts.* 50 (4): 15.

Thomas, B. Fossil Ape Skull Is a Game Ender. *Creation Science Update.* Posted on ICR.org September 17, 2019.

Tomkins, J. P. 2019. Recent Humans with Archaic Features Upend Evolution. *Acts & Facts.* 48 (4): 15.

Tomkins, J. P. 2019. *Homo erectus*: The Ape Man That Wasn't. *Acts & Facts*. 48 (10): 11-13.

Tomkins, J. P. 2020. *Homo naledi*: Another Failed Evolutionary Ape-Man. *Acts & Facts*. 49 (1): 10-13.

Tomkins, J. P. 2020. Human Chromosome 2 Fusion Never Happened. *Acts & Facts*. 49 (5): 16-19.

Tomkins, J. P. Human Genome 20th Anniversary—Junk DNA Hits the Trash. *Creation Science Update*. Posted on ICR.org April 12, 2021.

Tomkins, J. P. Human High-Altitude Habitation Reveals Adaptive Design. *Creation Science Update*. Posted on ICR.org March 22, 2021.

Tomkins, J. P. Humans and Neanderthals More Similar Than Polar and Brown Bears. *Creation Science Update*. Posted on ICR.org June 28. 2020.

Sherwin, F. 2017. Lucy Languishes as a Human-Ape Link. *Acts & Facts*. 46 (5): 10-13.

Sherwin, F. 2021. Man: Smart from the Start. *Acts & Facts*. 50 (8): 13.

## ABOUT THE INSTITUTE FOR CREATION RESEARCH

At the Institute for Creation Research, we want you to know God's Word can be trusted with everything it speaks about—from how and why we were made, to how the universe was formed, to how we can know God and receive all He has planned for us.

That's why ICR scientists have spent more than 50 years researching scientific evidence that refutes evolutionary philosophy and confirms the Bible's account of a recent and special creation. We regularly receive testimonies from around the world about how ICR's cutting-edge work has impacted thousands of people with God's creation truth.

## HOW CAN ICR HELP YOU?

You'll find faith-building science articles in *Acts & Facts*, our bimonthly science news magazine, and spiritual insight and encouragement from *Days of Praise*, our quarterly devotional booklet. Sign up for FREE at **ICR.org/subscriptions**.

Our radio programs, podcasts, online videos, and wide range of social media offerings will keep you up to date on the latest creation news and announcements. Get connected at **ICR.org**.

We offer creation science books, DVDs, and other resources for every age and stage at **ICR.org/store**.

Learn how you can attend or host a biblical creation event at **ICR.org/events**.

Discover how science confirms the Bible at our Dallas museum, the ICR Discovery Center. Plan your visit at **ICRdiscoverycenter.org**.

ICR
INSTITUTE
FOR CREATION
RESEARCH

P. O. Box 59029
Dallas, TX 75229
800.337.0375
ICR.org

# ICR MISSION STATEMENT

ICR exists to support the local church through….

### WORSHIP

- Glorify Jesus Christ by emphasizing in all ICR resources the credit He is due as Creator.
- Oppose the deification of nature by exposing Darwinian selectionism as an idolatrous worldview.

### EDIFICATION

- Help pastors lead, feed, and defend their flocks by providing scientific responses to secular attacks on the authority and authenticity of God's Word.
- Change Christians' view of biology by constructing an organism-focused theory of biological design that highlights Jesus' work as Creator.

### EVANGELISM

- Defend the gospel by showing how natural processes cannot explain the miracles in the Bible.
- Counter objections to the gospel by equipping believers with Scripture-affirming science.

# Chimps and Humans
## A Geneticist Discovers DNA Evidence That Challenges Evolution
### Dr. Jeffrey P. Tomkins

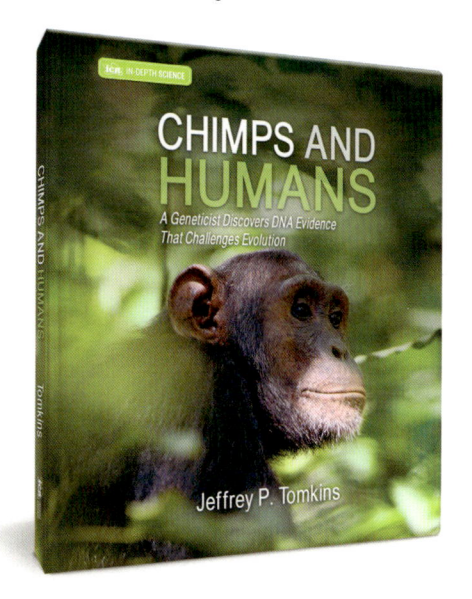

In *Chimps and Humans*, geneticist Dr. Jeffrey Tomkins dismantles evolutionary assertions of a close human-chimp relationship and shows they are too far apart to make a common evolutionary ancestor plausible. Christians have every reason to believe God created humans around 6,000 years ago.

Find out more about this book and other resources at **ICR.org/store**

**ICR**
INSTITUTE FOR CREATION RESEARCH